T0280983

Workload Automation Using HWA

With Architecture and Deployment Options

Navin Sabharwal
Subramani Kasiviswanathan

Apress®

Workload Automation Using HWA: With Architecture and Deployment Options

Navin Sabharwal
New Delhi, India

Subramani Kasiviswanathan
Chennai, 600127, India

ISBN-13 (pbk): 978-1-4842-8884-9
https://doi.org/10.1007/978-1-4842-8885-6

ISBN-13 (electronic): 978-1-4842-8885-6

Copyright © 2023 by Navin Sabharwal and Subramani Kasiviswanathan

This work is subject to copyright. All rights are reserved by the Publisher, whether the whole or part of the material is concerned, specifically the rights of translation, reprinting, reuse of illustrations, recitation, broadcasting, reproduction on microfilms or in any other physical way, and transmission or information storage and retrieval, electronic adaptation, computer software, or by similar or dissimilar methodology now known or hereafter developed.

Trademarked names, logos, and images may appear in this book. Rather than use a trademark symbol with every occurrence of a trademarked name, logo, or image we use the names, logos, and images only in an editorial fashion and to the benefit of the trademark owner, with no intention of infringement of the trademark.

The use in this publication of trade names, trademarks, service marks, and similar terms, even if they are not identified as such, is not to be taken as an expression of opinion as to whether or not they are subject to proprietary rights.

While the advice and information in this book are believed to be true and accurate at the date of publication, neither the authors nor the editors nor the publisher can accept any legal responsibility for any errors or omissions that may be made. The publisher makes no warranty, express or implied, with respect to the material contained herein.

Managing Director, Apress Media LLC: Welmoed Spahr
Acquisitions Editor: Aditee Mirashi
Development Editor: Laura Berendson
Coordinating Editor: Aditee Mirashi

Cover designed by eStudioCalamar

Cover image designed by Freepik (www.freepik.com)

Distributed to the book trade worldwide by Springer Science+Business Media New York, 1 New York Plaza, Suite 4600, New York, NY 10004-1562, USA. Phone 1-800-SPRINGER, fax (201) 348-4505, e-mail orders-ny@springer-sbm.com, or visit www.springeronline.com. Apress Media, LLC is a California LLC and the sole member (owner) is Springer Science + Business Media Finance Inc (SSBM Finance Inc). SSBM Finance Inc is a **Delaware** corporation.

For information on translations, please e-mail booktranslations@springernature.com; for reprint, paperback, or audio rights, please e-mail bookpermissions@springernature.com.

Apress titles may be purchased in bulk for academic, corporate, or promotional use. eBook versions and licenses are also available for most titles. For more information, reference our Print and eBook Bulk Sales web page at http://www.apress.com/bulk-sales.

Any source code or other supplementary material referenced by the author in this book is available to readers on GitHub via the book's product page, located at www.apress.com/978-1-4842-8884-9. For more detailed information, please visit http://www.apress.com/source-code.

Printed on acid-free paper

Thank you, goddess Saraswati, for guiding us to the path of knowledge and spirituality.

असतो मा साद गमय, तमसो मा ज्योतिर् गमय, मृत्योर मा अमृतम् गमय

(Asato Ma Sad Gamaya, Tamaso Ma Jyotir Gamaya, Mrityor Ma Amritam Gamaya)

Lead us from ignorance to truth, lead us from darkness to light, lead us from death to immortality.

Table of Contents

About the Authors... ix

About the Technical Reviewers xi

Acknowledgments xiii

Introduction xv

Chapter 1: Introduction to Workload Automation.................................. 1

Workload Automation Concepts 1

Introduction to HCL Workload Automation 6

HCL Workload Automation Strengths... 11

Summary.. 12

Chapter 2: HCL Workload Automation Architecture........................... 13

HWA Components... 13

HCL Workload Automation Components and Communication 16

Architecture Types 21

Summary.. 24

Chapter 3: HCL Workload Automation Deployments........................... 25

Planning Deployments 25

Stand-Alone Architecture 27

High Availability Architecture.. 32

Disaster Recovery Architecture 35

Containerized Deployment.. 39

HWA Deployment on Cloud 47

Summary.. 48

Chapter 4: Workload Design and Monitoring Using DWC.................................. **49**

Designing and Monitoring Workload Objects Using HWA Dynamic Workload Console 49

    Business Requirement... 55

    Summary... 96

Chapter 5: HWA for Managed File Transfers **97**

    Business Requirement... 97

    Summary... 109

Chapter 6: HWA Integration with SAP Application...................................... **111**

    Business Requirements 112

    Summary... 122

Chapter 7: Automate Job Executions on Microsoft SQL Server........................... **123**

    Business Requirement... 123

    Summary... 129

Chapter 8: Working with RESTful Web Services **131**

    Business Requirement... 132

    Summary... 138

Chapter 9: Submit, Orchestrate, and Monitor Jobs on a Kubernetes Cluster **139**

    Business Requirement... 139

    Summary... 144

Chapter 10: HWA Integration with Microsoft Azure.. **145**

    Business Requirementax 145

    Summary... 152

Chapter 11: HWA Integration with Ansible **153**

    Business Requirement... 153

    Summary... 159

Chapter 12: HWA Event Rule Management 161

HWA Integration with ServiceNow 161

Use Case 2: Auto-Remediation of Process Down 167

Summary 172

Chapter 13: Tool Administration and Best Practices 173

Daily Application Health Check 173

Housekeeping Procedure to Maintain Application Health 176

Database Maintenance Procedure 181

Database Backup and Restore Policies 183

Summary 183

Chapter 14: Alerting and Troubleshooting Issues 185

Configuring Alerts and Responses 185

Restore Services Post Outage 191

Troubleshooting Techniques 198

Summary 203

Chapter 15: HWA Reporting 205

Summary 210

Chapter 16: HWA Security 211

Traditional Model (File Based) 211

Role-Based Access Control 217

Summary 227

Index 229

About the Authors

Navin Sabharwal is Chief Architect and Head of Strategy for Autonomics at HCL Software. He is responsible for innovation, engineering, presales, and delivery of award-winning autonomics platforms. Navin has eight patent grants and has published numerous books on AIOps, DevSecOps, public cloud, automation, and machine learning. He can be contacted via his LinkedIn profile: www.linkedin.com/in/navinsabharwal/.

Subramani Kasiviswanathan is a Solution Architect, leading the Practice and Engineering Services for workload automation, AIOps, data analytics, and reporting in HCL DRYiCE, where he is responsible for supporting IP developments and service delivery in these areas. He has overall 19 years of IT experience and 6 years of experience in academics.

About the Technical Reviewers

Ummiti Ramesh is a Subject Matter Expert for workload automation solution and has overall 12 years of IT experience and 11 years of experience in HCL workload automation. Currently working as Technical Lead and Subject Matter Expert for all workload automation products supported under HCL DRYiCE, he is responsible for architecting solutions for greenfield customers and cross-tool migrations.

Franco Mossotto is a Sr. Software Architect responsible for HCL/IBM workload automation product family development. He joined IBM in 1998 as a Tivoli Workload Scheduler for z/OS developer. Franco has worked in design, development, and support of IBM Tivoli scheduling and provisioning products. In the scheduling area, he worked as a developer, chief designer, L3 technical leader for both Tivoli Workload Scheduler and Tivoli Workload Scheduler for z/OS, and as an architect for the development of cloud offerings. Since September 2016, with the IBM and HCL IP partnership, Franco moved to HCL Technologies with the rest of the team to continue his career with the HCL/IBM Workload Automation portfolio.

Acknowledgments

To my family for their love and support as always; without your blessings, nothing is possible. To my coauthors, contributors, and reviewers and the brilliant technical team Subramani Kasiviswanathan, Ummiti Ramesh, Franco Mossotto, and Ajmeer Mohideen – thank you for the hard work and quick turnarounds to deliver this book. It was an enriching experience, and I am looking forward to working with you again soon.

Thank you to Aditee and the entire team at Apress for making this happen.

Introduction

Workload automation has been the backbone of business processes. Most business processes involve job scheduling and workload automation to ensure that the systems are integrated and information flows across them in a seamless error-free fashion. In the space of workload automation, there have been so many advancements in terms of integrating your complex workload, workflow, and business processes across automation platforms, ERP systems, and business applications from mainframe to multicloud. Unlike other systems of automation, workload automation is focused on real-time processing, predefined event-driven triggers, and situational dependencies. Workload automation offers centralized control of the management of multiple tasks, making it possible to schedule enterprise-wide tasks. Further, it supports the timely completion of tasks. WLA increases efficiency, reduces the turnaround time for workflows, and reduces errors in end-to-end processes.

The book discusses how HCL workload automation solution has been able to meet automation requirements of the digitally transformed platforms. It specifies detailed architecture and deployment options. The course demonstrates best practices to deliver robust workload automation solution by providing insights into best practices and integration with various systems.

The book can be a reference material for beginners as well as advanced users of workload automation. It not only provides information on how to use the tool but also gives ideas on how to choose the right candidates or use cases for workload automation. The course contains numerous use cases and their implementation procedures that can be referred for any workload automation requirements.

CHAPTER 1

Introduction to Workload Automation

This chapter introduces workload automation concepts, mandatory features that a workload automation solution should possess, the components of HCL workload automation, network communication between various components, different processes that constitute HCL workload automation service, and the strengths of HCL workload automation.

Note As a takeaway of reading this chapter, the reader is expected to understand the concepts of workload automation and the need for this solution in IT operations.

Workload Automation Concepts

Workload automation is not new. Gartner introduced the concept of workload automation in their report "Hype Cycle for IT Operations Management" during the year 2005. IT operation teams were following the traditional approach of static and manual job scheduling. Before the evolution of workload automation tools, job scheduling and maintenance was a manual and error-prone task. With the arrival of job scheduling tools, this was automated and made easier through usage of standardized tools.

Workload automation addresses the requirements of automating the executions of business processes and transactions. It is also known as a job scheduling system. Workload automation solution automates the workflows which can be executed at a particular time, or they can be executed dynamically based on events and triggers. The following Figure 1-1 depicts a typical stand-alone job stream and the various objects that constitute a job stream, which we will go over in the upcoming sections and chapters.

1

© Navin Sabharwal and Subramani Kasiviswanathan 2023
N. Sabharwal and S. Kasiviswanathan, *Workload Automation Using HWA*,
https://doi.org/10.1007/978-1-4842-8885-6_1

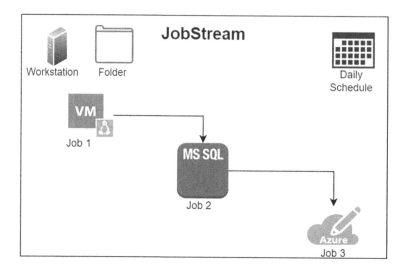

Figure 1-1. *HCL Workload Automation Stand-Alone Job Stream*

Any workload automation solution should be able to address the following requirements:

- Centralized job management

  Every application/technology has a built-in scheduler that is natively provided and supported by the vendor, for example, Informatica offers Datastage, MSSQL offers SQL scheduler, Windows offer Windows scheduler, Unix offers Crontabs, SAP offers SAP scheduler, etc. The native schedulers can automate the business processes and transactions pertaining to their application. These are aware of the dependencies within their application; however, the limitation is that these schedulers are not aware of the status of the processes running for other applications on which they are dependent. In a distributed environment where applications are loosely coupled and dependent on each other, it becomes highly important to visualize the status of all business processes centrally in a unified console to get a holistic view of what is going on with all the business processes running across the application landscape. A business process may cut across multiple transactions which are happening in different applications, and multiple jobs may need to be triggered in a defined sequence across multiple applications for a business process to execute. A job scheduling software helps in automating this by providing a master scheduler which can control the execution of jobs across different applications and platforms. Thus, from an IT operations perspective, the operators and administrators can get a central console to view the status of these connected or dependent jobs as well as manage the complex scheduling and execution of these jobs in an easy-to-use console. Figure 1-2 depicts a typical job stream with external dependencies and the various objects that constitute it.

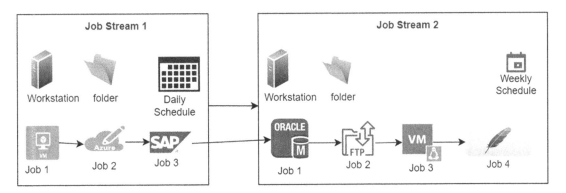

Figure 1-2. *HCL Workload Automation Interdependent Job Streams*

- Time-based and event-driven scheduling

  The workload automation solution should be able to schedule the business processes to run on the target based on a calendar/date/time. It should also be able to trigger business processes/workflows based on certain conditions or events that are occurring on the target machines.

- Application and platform integrations

  In a traditional approach of handling business processes and their dependencies, a scheduler or the application developer had to write wrapper scripts using any scripting languages like PowerShell, Shell, Perl, Python, VBScript, etc., to provide all inputs for the job to trigger on the target application server like a valid application user and their credentials, application profiles, source path and job to trigger, database details, queries, procedures, and so on. In a transformed workload automation environment, we have the benefits of

3

application and technology adaptors or plug-ins that can help to seamlessly integrate with the applications or platforms and provide an interface for all the inputs needed for batch processing on target applications and platforms. This eliminates the need for custom development and usage of error-prone scripts for automating job scheduling. A large repository of adaptors and plug-ins provide out-of-the-box integration with widely used applications and platforms.

- Ability to handle/store outputs in variables

  The business processes run on the set targets generate the output or log files that are stored either on the server or client or both. The workload automation solution should be able to scan the output, trace the required messages, and store them into variables. These variables can then be used for further batch processing. This enables the workflow to proceed based on output provided by each step that acts as an input to the next step.

- Conditional batch processing

  The related business processes are grouped and defined as one batch. They are executed according to the sequence based on the dependencies set on them. The dependencies can be within the same batch or a different batch. The dependencies within the same batch are called internal dependencies, and the ones from other batches are called external dependencies. The batches or any individual business processes can be set to execute based on certain logical conditions. The values that are stored in variables after an output scan can be compared, and batches can be set to trigger based on the results of the comparison.

- Handling resources virtual and physical

  It is very important to ensure that acceptable workload is submitted to execute on the target hosts at any given point in time so that it does not impact the performance of the application and the databases. Any workload automation solution should have the ability to manage the workload by configuring physical or virtual resources to control the load that is exerted on the target.

- Ability to manage alerts, events, and logs

  Workload automation solution should be able to configure and generate the alerts for events such as a status change, process running longer than expected, possible impact in meeting the SLA, service/process failure, etc., and it should also be able to log the details of the events or errors that will help schedulers and application users to troubleshoot the issues and figure out the root cause. They should also be able to capture the application logs and store them on the server or client or both.

- Ability to provide self-service and RBAC

  It is an expectation from the workload automation solution to provide the users with self-service capabilities where users can view, edit, execute, and manage the business processes pertaining to their application only. The access granted should be granular and based on the roles.

- Manage releases for process/schedule definitions

  Just like any other development environment, workload automation jobs are also created in a development environment and moved to testing. The tools should provide capabilities to promote the preproduction environment to production environment.

- Critical path and batch impact analysis

  The solution should possess the capability to define critical processes in a batch or workflow and should be able to proactively identify potential delays, likely errors and how they could impact systems and business users. They should also be able to alert such cases to schedulers and application users in a timely fashion so that actions can be taken proactively so as to avoid or minimize disruptions to the critical business processes.

- Application-level high availability and disaster recovery

  Workload automation architecture should provide capabilities for application-level high availability, i.e., when primary node of an application is not in service, the architecture should support automatic failover to a secondary node of the application and restore the service with minimal or no downtime. Similarly, it should also provide the built-in mechanism to restore the services on primary once it is up and running. The application architecture should also address the requirements of disaster recovery at times when the data center itself is down and services need to be restored from a DR location.

- Security

  The solution must be able to manage security at all levels, i.e., user-level security, application security, and data security.

Introduction to HCL Workload Automation

HCL workload automation is an enterprise solution to meet hybrid workload management needs of modern enterprises. It provides a platform for centralized job management across different platforms and application environments via its strong integration capabilities. It caters to all KPIs that a workload automation solution should meet as described in the previous section.

The solution has three main components in the portfolio:

1. HCL workload automation for Z

2. HCL workload automation for distributed

3. Dynamic Workload Console (DWC)

HCL workload automation comprises of the following components:

- HCL workload automation engine

  This is a scheduling engine which is the heart of the system that governs and processes the tasks pertaining to the scheduler. It is assigned with specific roles during the installation of the software in the HCL workload automation network. We will see more in detail about the roles in upcoming chapters. Figure 1-3 depicts the core components of HCL workload automation.

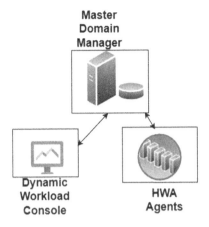

Figure 1-3. *HCL Workload Automation Components*

- Dynamic workload console

  DWC is a unified and centralized web-based graphical user interface for HCL workload automation. This component provides a user-friendly interface to access all the workload automation functions. The following Figure 1-4 depicts Dynamic Workload Console (DWC).

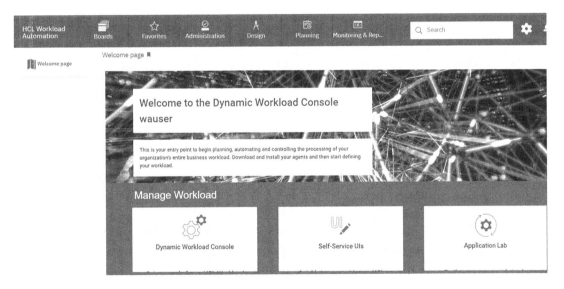

Figure 1-4. *Dynamic Workload Console*

- HWA agents

  Agents are HWA software that are installed on the end points where the workload automation needs to be established. These end points can be any application servers, database servers, web servers, etc. This component helps execute jobs on the intended end points and send back the output and status to the master.

  A typical/basic HCL workload automation network is shown in the following illustration. We will discuss each of these elements in detail.

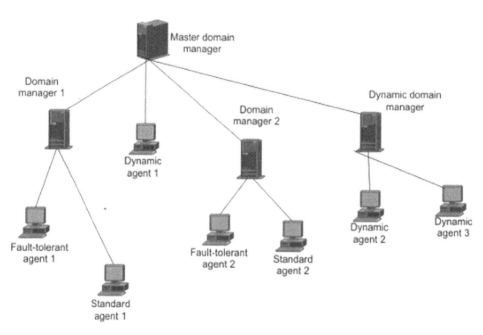

Figure 1-5. *HCL Workload Automation Network*

- Master domain manager

  This is the central processing unit of a workload automation system and is a management hub. Master domain manager communicates directly with the database to collect information related to HWA objects and generate the production plan that needs to be supplied to all its domain managers and fault tolerant agents. All the other components like the domain managers, agents, etc. receive instructions from the master domain manager as to what is their schedule/tasks for a defined business day.

- Domain manager

  It plays an integral part in a widely distributed and large network where the workload is required to be divided and managed locally in smaller groups to ensure balanced workload and smooth execution of batches.

- Dynamic workload broker

  This component governs and provides all services related to the dynamic scheduling which consists of dynamic agent, pool, dynamic pool, etc.

- Dynamic agent

  This component is installed on the target machines or application servers on which the execution tasks need to happen. Agents are automatically discovered by their domain or master domain managers once they are installed in the network and get registered with them. They then start receiving instructions from their domain or master domain managers as to what to process and when to process.

- Fault tolerant agent (FTA)

  FTA is installed on the target machines or application servers on which the tasks need to be executed. As the FTAs receive a copy of their schedule well in advance from the domain manager or master domain manager, they are aware of their dependencies and hence can provide scheduling service even when their domain or master domain manager is not active in the network.

- Pool

  Pool is a virtual definition that comprises of one or more dynamic agents. This component performs load balancing and launches the jobs on the targets based on the resource availability.

- Dynamic pool

  DP is a virtual definition that comprises of one or more dynamic agents. This component performs load balancing and launches the jobs on the targets based on the resource availability. The user has the advantage to control the target where the jobs need to be executed. Moreover, the processes can be executed based on the availability of physical resources such as CPU, disk, etc.

- Standard agent (SA)

  This agent type executes instructions as directed by its manager.

- Extended agent (XA)

  This is governed by FTA but also extends additional attributes to execute workload for specific applications like PeopleSoft, Oracle applications, SAP R3, etc.

After having gone through the HWA components and its functions, let us understand some basic concepts and terminologies used in HWA.

Production Plan (JnextPlan)

JNextPlan is a script that runs every day in HWA during the start of a defined business day and follows a sequence of steps to move the old plan and generate the new schedule/tasks for the current business day, as illustrated in Figure 1-6. The production plan generated for the current day is called the "Symphony file."

The following are the steps involved during the generation of a production plan in HCL workload automation.

StartAppServer

This job checks the availability of the application server and restarts it in case it's not running.

MakePlan

Master domain manager connects to database via the web server (Liberty) and creates the production plan based on the scheduling conditions and dependencies defined in its database. During the plan creation, it locks the database using "Planner process" to avoid any new processing during this time that can obstruct the plan generation process. This process creates a pre-production file known as "SymNew."

- SwitchPlan

  "Planner process" then updates the pre-production plan and
  "Stageman" merges pre-production file SymNew with the pending
  process executions and incomplete schedules that need to be
  carried forward from previous day and consolidates it in a file called
  "Sinfonia"; a copy of it will then be updated to the new production
  plan file called Symphony for the current day. The Symphony file for
  the previous day will get archived in Schedlog folder. The planner
  then unlocks the database and makes it available for use. The newly
  created symphony file gets distributed to its agents.

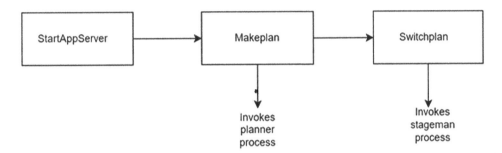

Figure 1-6. *JnextPlan Flowchart*

HCL Workload Automation Strengths

WA provides extensive capabilities and features for job scheduling
and orchestration. Let us go through some of the key capabilities that
HWA provides.

- HWA is evolving to become a more general orchestrator tool with
  extensive scheduling capabilities.

- HWA provides UI and REST API import/export features to allow the
  integration of HWA in CI/CD pipeline. You can use this feature in
  your pipeline for running any scheduled jobs as a part of the testing
  process.

- HWA has a development toolkit to build new job types and many out-
  of-the-box integrations. The new Automation Hub website provides

information and plug-ins to orchestrate multiple automation domains.

- HWA Web Console is highly intuitive and provides effective graphical views and dashboards with alerts on KPIs. The Self-service Catalog provides mobile access for business users.

- All HWA components (Server, Console, and Agents) can be deployed on containers. The Kubernetes job type allows to manage containerized applications.

- HWA provides full REST API support (anything that can be done with the Console), and a CLI is available to interact with the product.

- HWA provides versioning for all the database objects with compare and restore capabilities. All actions are logged and available for auditing.

- HWA allows SLA management and predictive scheduling through the Workload Service Assurance and the What-If features.

Summary

To summarize the content of this chapter, the reader should have understood the following:

- Workload automation concepts

- Various HCL workload automation terminologies and their definitions

- HWA JNextDay plan

- HWA strengths

CHAPTER 2

HCL Workload Automation Architecture

In this chapter, we will cover details on deployment architecture of HWA. The objective is to impart, among the readers, a clear understanding of the architecture components and communication between them.

Note The readers are expected to have read and understood Chapter 1 so that they can relate with and cover Chapter 2.

HWA Components

The following are the HCL workload automation components:

HCL Workload Automation Master Engine/Scheduler

The HWA master is called the scheduler. This engine is the application server component that processes all the instructions that are scheduled to be executed. It plans the schedules day wise and distributes the workload to its agents to process as per definitions, hence governing all the processes that are scheduled to run for a particular business day and updating their status to its database.

HWA extends its service in the following areas:

- The schedule plan is generated by the master for the business day; previous plans are captured in the server log location for further reference/troubleshooting.

© Navin Sabharwal and Subramani Kasiviswanathan 2023
N. Sabharwal and S. Kasiviswanathan, *Workload Automation Using HWA*,
https://doi.org/10.1007/978-1-4842-8885-6_2

HWADATA/schedlog

- The master captures audit information for every access/change to scheduling definition or the plan and store them in the database. The same information is also logged in log files on the master and FTA where the command is run.

HWADATA/audit

- The master and the agents capture the logs of all the business processes that are scheduled to run on the remote systems. This shall be used for troubleshooting and audit purposes.

- The master captures the logs of all the communications that happen between HWA components internally and stores it in the log location for further reference/troubleshooting.

HWADATA/stdlist/traces/<DATE>_TWSMERGE.log

- The master provides options to set up granular role-based access controls (RBAC) to its local or domain users.

- The architecture supports various databases to which the master can communicate with.

- The HCL workload automation integrates with various application/technology/platform/server environments to directly interface with the application to execute/monitor/schedule/manage via plug-ins.

- The master has Application HA capability to fail over and fail back the services automatically between the master and the backup master to ensure service maximum availability.

- Dynamic Workload Console

  - DWC is the front-end HWA console which brings in the feature of graphical user interface for the users to monitor, design, and manage their workload. DWC also provides the functionality to define and administer HWA and its components.

- DWC provides functionality to restrict object-level access to users of various roles like administrator, operator, developer, etc. so that they can see specific views according to their roles.

- DWC provides functionality to design specific dashboards and insights about the various business processes for an application, system availability, critical jobs, min/max duration, reports based on job status, etc.

- Fault Tolerant Agent

  - FTA provides functionality to run command jobs or executables on the remote machines where FTA agent is installed.

  - FTA provides functionality to perform agentless job executions on a remote host via the SSH protocols.

  - FTA provides functionality to run business processes on a remote host even when the master that it reports to is not in service as it is well aware of its workload via the plan circulated by its master.

- Dynamic Agent

  - DA provides the functionality to run jobs for advanced job types like RESTAPI jobs, Database, Cloud Native, etc., via job templates. Unlike FTAs, plan is not delivered to DAs by master; hence, they receive the instructions in real time from their master about their workload that they need to run. They can also provide the same feature like FTA.

  - The remote servers where the business processes are launched can be grouped in a single dynamic pool on which the processes can be submitted when specific conditions satisfy like available CPU, memory, etc.

  - DA provides the functionality of the gateway server which receives connections from various dynamic agents and validates and approves them before sending the communication to the target system component.

- Dynamic Workload Broker

  - DWB enables the dynamic agent to plan and run the workload on specific remote servers or static/dynamic pool after analyzing the job definitions, resources, etc.

  - DWB job dispatcher dispatches and submits the jobs that it receives from the master to the respective dynamic agents to process. Before submitting the resource, advisory analyzes and decides the static/dynamic pool on which the job has to be submitted.

- Liberty

  - Liberty is an application server runtime environment that helps in communicating between various components of HWA including the database.

  - On distributed platforms, Liberty provides both a development and an operational environment.

- Database

  - Database is a data store for HWA objects. HWA architecture supports various databases to which the master can communicate with. Some of the supported databases are HCL OneDB, Informix, Azure SQL database, DB2, MSSQL, and Oracle.

  - DB2 database is bundled with HWA package and does not need any separate licenses. HA and DR functionality is also inclusive.

HCL Workload Automation Components and Communication

HCL workload automation has several internal processes running that help in segregating and manage networking, resolving dependencies, and running the jobs on the target systems. It performs these activities with the help of message queues. There are many such processes that get initiated during the HWA application startup.

The diagrams shown in Figures 2-1 and 2-2 depict the processes that constitute the internal communication between the various components of HCL workload automation.

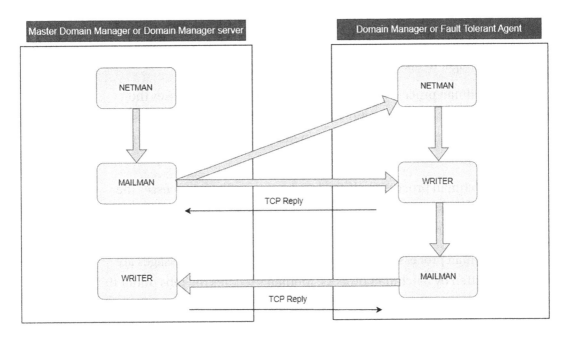

Figure 2-1. *HCL Workload Automation Interprocess Communication*

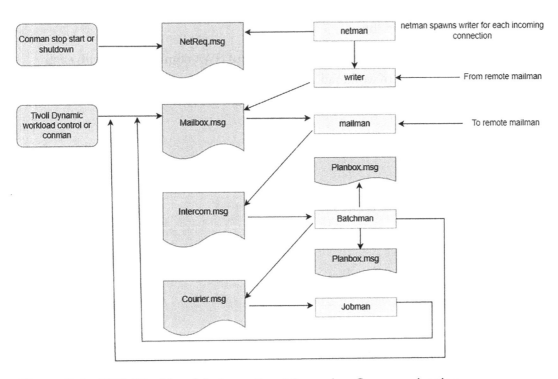

Figure 2-2. *HCL Workload Automation Messaging Communication*

- Netman process reads NetReq.msg file and receives all different network communications between the FTA agents and domain manager to take respective actions.

- Mailman process reads the Mailbox.msg file. It processes the messages and instructions from the message file. These messages are coming from all the components that constitute HCL workload automation.

- Batchman process reads the Intercom.msg file. It processes the messages and instructions from the message file. These messages are coming from local mailman process.

- Jobman process reads the Courier.msg file. These messages are written by batchman process. Jobman executes the jobs under the direction of batchman process. Jobman process is owned by root user to enable it to execute the job by any user.

- Batchman process writes the messages to Planbox.msg and Server.msg files. These instructions are read by the HWA scheduling engine for further processing.

This part explains how the various components of HWA communicate with each other to serve the purpose of workload automation.

Figure 2-3 shows the detailed communication path of HWA component.

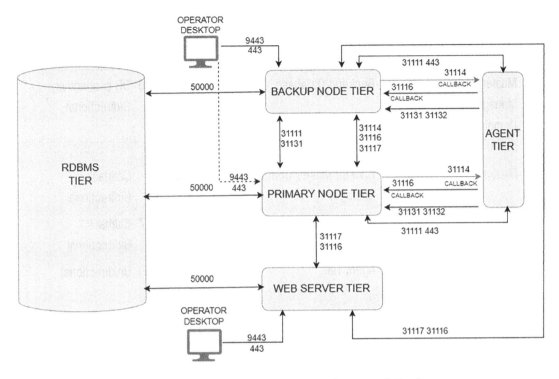

Figure 2-3. *HWA Components Communication/Network Path*

- RDBMS Tier denotes the database component; from Figure 2-3, it connects to DB2 database, and default DB2 database port is 50000. In case of other databases, the port numbers will change.

- Application Server Tier talks about the HWA master and Backup master domain managers.

- Web server tier is the DWC Dynamic Workload Console.

- Agent Tier is represented by the fault tolerant agent, dynamic agent, extended agent, and standard agent.

Table 2-1 provides a list of default ports that are used for communication among the HWA components.

Table 2-1. *List of HWA communication ports*

S.no	Source	Destination	Port	Rule notes
1	Master Domain ManagerBackup Master Domain ManagerWeb server	Backend Database	500005000050000	DB connectivity Bidirectional
2	Master Domain Manager	Backup Master Domain Manager	3111131131	Callback Bidirectional
3	Master Domain Manager	Backup Master Domain Manager	311143111631117	Callback Bidirectional
4	Master Domain ManagerBackup Master Domain Manager	Agent Tier	31114	Unidirectional
5	Agent Tier	Master Domain ManagerBackup Master Domain Manager	3113131116	Unidirectional
6	Agent Tier	Master Domain ManagerBackup Master Domain Manager	31111443	Bidirectional(HTTPS)
7	Operator Desktop	Master Domain ManagerBackup Master Domain Manager	9443443	Unidirectional
8	Operator Desktop	Web Server	9443443	Unidirectional
9	Master Domain ManagerBackup Master Domain Manager	Web Server	3111631117	Bidirectional

Architecture Types

This part explains the various HWA deployment architecture options. The following are the types of architectures that can be used based on the environment size and complexity.

- Stand-alone

    Stand-alone architecture is proposed for small customers with minimum number of jobs. Typically, less than 500 jobs would fall under this category. Along with number of jobs, another factor is the complexity of the environment and jobs. This architecture should be used for deployments where the complexity is minimal. Stand-alone architecture comprises of a single Master Domain Manager, DWC, Liberty, and Database that manage the workload. This architecture choice introduces a downtime when the master services are stopped due to some reason, and this could cause a potential business impact. Figure 2-4 represents a typical stand-alone deployment architecture.

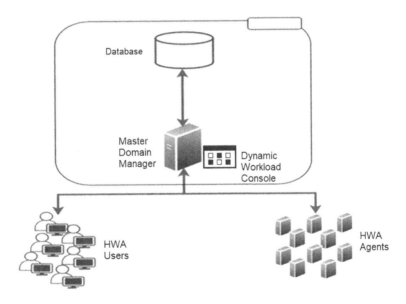

Figure 2-4. *Stand-Alone HWA Architecture*

- High availability

  High availability architecture is proposed for medium and large customers with more than 500 jobs and higher complexity of jobs. High availability architecture comprises of Primary and Backup Domain Managers with all components highly available. This avoids downtime when the master services are stopped on any of the active master domain managers. It introduces high availability for both HWA and database components. In this architecture, we will have two master domain managers and consoles. We would need to make sure network communication is established between these two servers as well as agents connected to the master servers. The services on primary master will be active while the services on the backup master will be passive. Whenever the primary master server goes down, HWA performs the health check and identifies that the primary master is down and initiates the auto failover option so that the backup master will take the role of primary master server and inform the agents connected to the primary master server to connect with new primary master server (backup master). The advantage is that during the failover, there will not be any issues with job processing and no downtime is required. Once the primary master domain manager is active, HWA has the functionality to perform failback from backup master to primary master. However, this process needs to be initiated manually after required health checks are performed on primary master. This process also does not impact job processing or introduce any downtime.

Note   The HA architecture should be deployed in a scenario where we need the services to be available all the time within the data center.

Figure 2-5 represents a typical high availability deployment architecture.

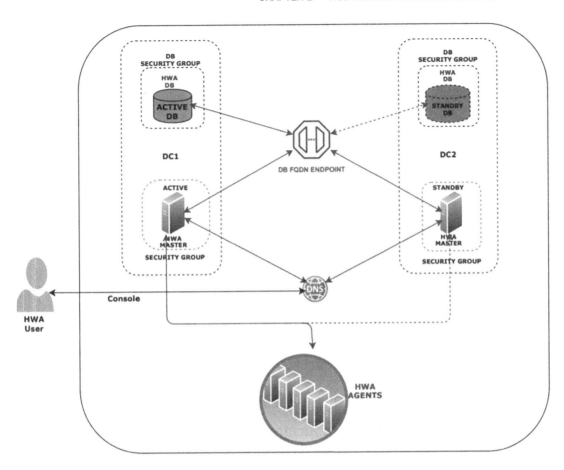

Figure 2-5. *High Availability HWA Architecture*

- Disaster recovery

  Disaster recovery architecture is similar to the high availability architecture but is proposed for customers who would like to have their DR set up in a different location to cater to disaster scenarios. This setup ensures high availability of the HWA services within the data center as well as ensures services to failover and be available at a DR data center when the primary data center is down or unavailable. The DR master server takes the role of a backup master but is put to use only when the backup master in the primary data center is also not available. This means that the data is in sync between the primary, backup, and the DR servers.

It takes approximately 5–10 minutes for the DR process to be triggered and bring up all the services on the DR site. We will be covering this scenario in the upcoming chapters.

Figure 2-6 represents a typical disaster recovery deployment architecture.

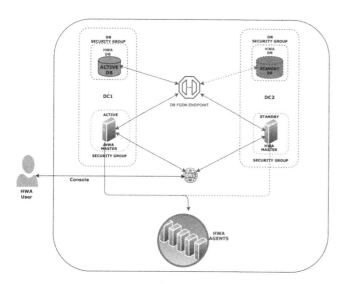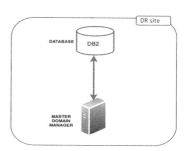

Figure 2-6. *Disaster Recovery HWA Architecture*

Note   The DR architecture should be deployed in a scenario where we need the services to be available at a DR data center when the primary data center is down or unavailable.

Summary

To summarize the content of this chapter, the reader should have understood the following:

- Workload automation components and their functions in detail

- HCL workload automation network communication

- HWA architecture types and their features

## CHAPTER 3

HCL Workload Automation Deployments

HCL workload automation provides multiple deployment options to their customers according to the business requirements and the criticality. The following are the various deployment options and recommendations. In this chapter, we will cover these architecture options in detail. This will allow customers to decide on a suitable architecture for their enterprise requirements.

- Stand-alone architecture

- High availability architecture

- Disaster recovery architecture

- HWA deployment on containers

Note   This chapter is an extension from the previous chapter to provide detailed information on the deployment options.

Planning Deployments

While planning for any deployment, prerequisites are important and so is the case with HWA. Prerequisites must be completed before the deployment.

- HWA requires minimum of 2 GB of virtual swap memory space. Figure 3-1 provides the memory requirements for each component in MB.

© Navin Sabharwal and Subramani Kasiviswanathan 2023
N. Sabharwal and S. Kasiviswanathan, *Workload Automation Using HWA*,
https://doi.org/10.1007/978-1-4842-8885-6_3

Memory	MDM or DDM	Dynamic agent or z-centric agent	Fault-tolerant agent	File Proxy
Recommended	8192	1024	1024	2000
Required	4096	1024	512	1000

Figure 3-1. *Memory Requirements*

- Temporary directory must have a maximum of 1GB free space.

- The disk space for HCL workload automation is specified based on the data, log files, etc.

- HCL workload automation should have a dedicated user to install the HWA application. HWA Windows user passwords can include any alphanumeric characters and ()!?=^*/~[]$_+;:.,@`-#. And for Unix user, passwords can include any alphanumeric characters and ()!?=*~_+.-.

- Database must be installed prior to the HWA implementation. Even if not mandatory, for the rest of the chapter, we have assumed to have two database instances (db2inst1 and db2inst2) created on db server. db2inst1 will be used for HWA database, and db2inst2 will be used for DWC database. Make sure the database user should have full privileges to read/write the information into the database. In this chapter we are considering DB2 as the backend database for HWA.

- Communication between HWA engine and database has to be enabled. Please work with your network team to enable the required ports. Refer to Chapter 2 for the network communication in Figure 2-2. Test your port connectivity between the server where you are planning to install the HWA engine and the database servers.

- Once the preceding components are configured, we can proceed with HCL workload automation deployment.

Stand-Alone Architecture

This type of architecture is proposed when the customer environment is small and less critical workload. In this architecture, there is no high availability; thus, if the system goes down you will have to use other techniques at the virtualization level to bring up the machine or restore from a backup to bring up the machine. Database must be installed prior to HWA installation. In this deployment, we will install Liberty, MDM (Master Domain Manager), and DWC (Dynamic Workload Console). Figure 3-2 depicts a typical stand-alone HWA setup.

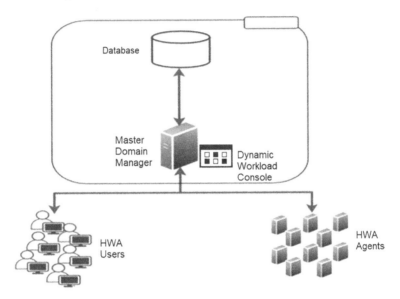

Figure 3-2. *Stand-Alone Architecture*

The following is the deployment procedure:

- Download the Liberty, HWA, and DWC and db2 package from Flexnet portal. Following is the link to the portal.

  ```
  https://id.hcltechsw.com/login/login.htm?
  ```

- Install the Liberty using the following command:

  ```
  java -jar /image/wlp-base-all-20.0.0.11.jar --acceptLicense
  <inst_dir>
  ```

- Create the database using create_database.sql query. Please note the database to be created using database instance user, so switch to db2inst1 user and run create_database.sql using the following command.

```
db2 -tvf create_database.sql
```

- Once the database is created, switch to root user and configure the database using configureDb.sh Script like the following:

```
 ./configureDb.sh --wlpdir <inst_dir> --rdbmstype DB2 --
dbhostname <dbhostname> --dbport 50000 --dbname
TWS --dbuser <db2user> --dbadminuser <db2adminuser> --
dbadminuserpw <XXX>
```

- Once database is configured, install the HWA application using serverinst.sh script.

```
./serverinst.sh --acceptlicense yes --rdbmstype DB2 --
dbhostname <dbhostname> --dbport 50000 --dbname TWS --
dbuser <db2user> --dbpassword <XXXX> --wauser wauser --
wapassword <XXXX> --wlpdir <inst_dir> --inst_dir <inst_dir> --
HCL --thiscpu <CPUNAME> --brwksname <DWB> --
xaname <XA> --displayname <DA> --componenttype MDM --
hostname <serverFQDN> --licenseserverid <XXXXX>
```

- Once HWA is installed, move to install DWC. Create the DWC database using create_database.sql query. Please note database to be created using DWC instance user db2inst2, so switch to db2inst2 user and run create_database.sql as the following to create the database.

```
db2 -tvf create_database.sql
```

- Once the database is created, switch to root user and configure the database using configureDb.sh script like the following:

```
./configureDb.sh --wlpdir <inst_dir> --rdbmstype DB2 --
dbhostname <dbhostname> --dbport 50001 --dbname
DWC --dbuser <db2user> --dbadminuser <db2adminuser> --
dbadminuserpw <XXX>
```

- Install DWC using dwcinst.sh script.

```
./dwcinst.sh --acceptlicense yes --rdbmstype DB2 --user
wauser --password <xxx> --dbname DWC --dbuser <dbinst2> --
dbpassword <xxxx> --dbhostname <dbhostname> --dbport
50001 --wlpdir <inst_dir> --inst_dir </inst_dir/DWC/>
```

- Login to DWC portal and validate the environment.

    Login to DWC using wauser and connect to DWC portal using this
    link `https://<hostname/ipaddress>:9443/console/login.jsp`.

- To connect DWC to HWA server, navigate to Administator➤Manage
  Engine. Refer to Figure 3-3.

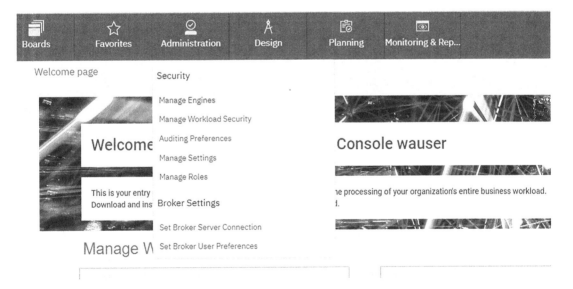

Figure 3-3. *Manage Enginetion Page*

- This opens the interface to create the new engine. Refer to Figure 3-4.

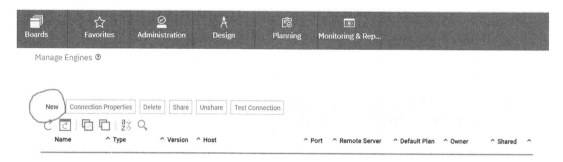

Figure 3-4. *New Engine Creation Page*

- This opens the interface where we update HWA credentials and database credentials. Refer to Figure 3-5. Select "Test Connection."

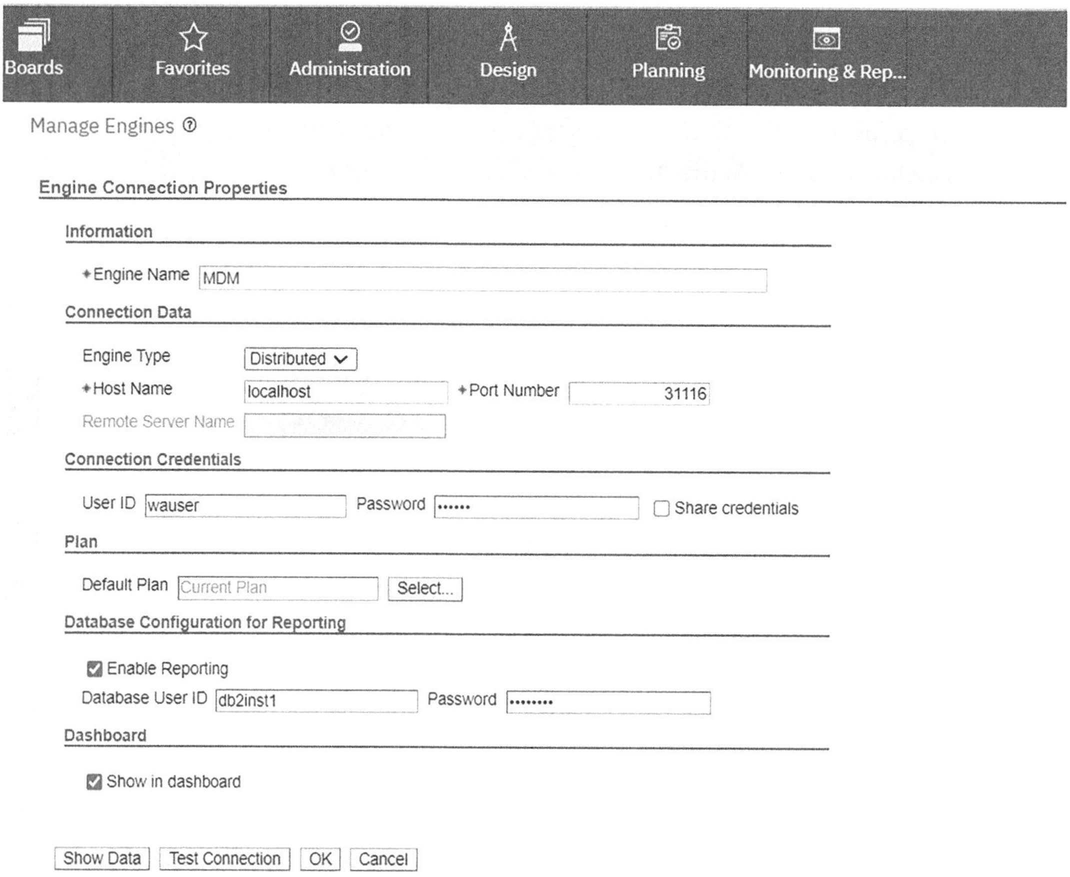

Boards | Favorites | Administration | Design | Planning | Monitoring & Rep...

Manage Engines ⑦

Engine Connection Properties

Information

✦Engine Name   MDM

Connection Data

Engine Type   Distributed ⌄

✦Host Name   localhost   ✦Port Number   31116

Remote Server Name

Connection Credentials

User ID wauser   Password •••••   ☐ Share credentials

Plan

Default Plan Current Plan   Select...

Database Configuration for Reporting

☑ Enable Reporting
Database User ID db2inst1   Password ••••••••

Dashboard

☑ Show in dashboard

Show Data | Test Connection | OK | Cancel

Figure 3-5. *New Engine Properties*

- This opens the interface where it will show the status of the connection to the database. Refer to Figure 3-6.

Figure 3-6. *Test Connection Results*

- We can see that connection to the database is successful, and to do further validation, we can now try to open the manage workload definition page. Refer to Figure 3-7.

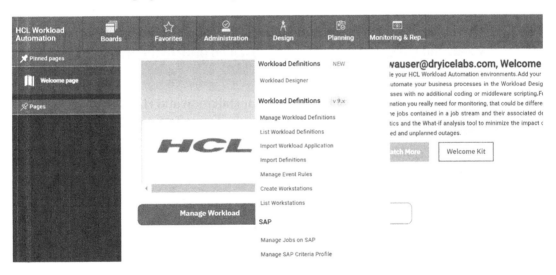

Figure 3-7. *Landing Page of DWC*

- This opens interface where we can confirm that the database connection to the HWA engine is working fine from console. Please refer Figure 3-8.

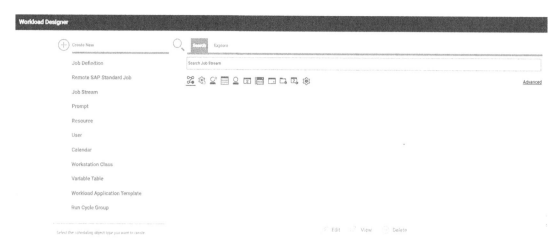

Figure 3-8. *Workload Designer Page*

High Availability Architecture

- High availability architecture is proposed when the customer is having >5000 and <10000 jobs that are critical. In this architecture, we will have primary and secondary servers installed with Liberty, MDM, and DWC. Database has to be installed prior to the HWA implementation on both the servers.

- Please refer the following links for DB2 installation and HADR configuration:

  www.ibm.com/docs/en/db2/11.5?topic=servers-db2-installation-methods

  www.ibm.com/support/pages/step-step-procedure-set-hadr-replication-between-db2-databases

  Figure 3-9 depicts a typical high availability HWA setup.

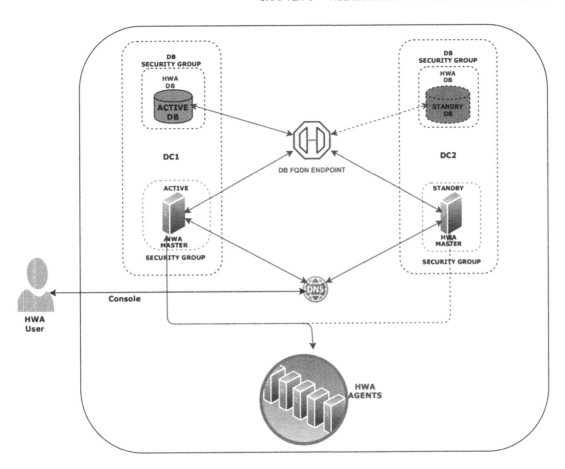

Figure 3-9. *High Availability HWA Architecture*

The following is the deployment procedure:

- Download the Liberty, HWA, and DWC and db2 package from Flexnet portal.

- Communication between primary and secondary servers of HWA and database servers must be enabled.

- Perform the same steps as stand-alone architecture for installing Liberty, HWA, and DWC.

- Configure HADR on db2 databases for both the primary and secondary servers.

- On the primary, run the following commands to enable the database TWS in high availability mode:

```
db2 update db cfg for TWS using HADR_LOCAL_HOST
<Primary server>
db2 update db cfg for TWS using HADR_LOCAL_SVC 60001
db2 update db cfg for TWS using HADR_REMOTE_HOST
<Secondary server>
db2 update db cfg for TWS using HADR_REMOTE_SVC 60001
db2 update db cfg for TWS using HADR_REMOTE_INST db2inst1
db2 update db cfg for TWS using LOGINDEXBUILD ON
```

- On the primary, run the following commands to enable the database DWC in high availability mode:

```
db2 update db cfg for DWC using HADR_LOCAL_HOST
<Primary server>
db2 update db cfg for DWC using HADR_LOCAL_SVC 60002
db2 update db cfg for DWC using HADR_REMOTE_HOST
<Secondary server>
db2 update db cfg for DWC using HADR_REMOTE_SVC 60002
db2 update db cfg for DWC using HADR_REMOTE_INST db2inst2
db2 update db cfg for DWC using LOGINDEXBUILD ON
```

Note   Run the same commands on the secondary database server, and ensure changing the HADR_LOCAL_HOST and HADR_REMOTE_HOST names.

- To start the HADR service on the databases, run the following commands sequentially, first on primary database server and next on secondary database server.

Primary:

```
db2 start hadr on database TWS as primary
db2 start hadr on database DWC as primary
```

Secondary:

```
db2 start hadr on database TWS as standby
db2 start hadr on database DWC as standby
```

- To validate if the HADR role has been activated on the primary and secondary servers, run the following command sequentially, first on primary database server and next on secondary database server.

```
db2pd -db TWS -hadr
db2pd -db DWC -hadr
```

In the output, the following parameters should have values:

```
HADR_ROLE = PRIMARY
HADR_SYNCMODE = NEARSYNC
HADR_STATE = PEER
HADR_CONNECT_STATUS = CONNECTED
```

- For auto failover enAutomaticFailover option has to be set to "Yes" in HWA end using the optman command as the following. By default, this is set to "Yes." Use "optman show af" to show the current value, and use "optman chg af=yes" to set it to yes. The file path is <TWSDATA/mozart>.

```
enAutomaticFailover / af = YES
```

- Perform the validation steps same as stand-alone architecture.

Disaster Recovery Architecture

- Disaster recovery architecture is same as high availability architecture; this architecture is proposed when you require a disaster recover setup. In this architecture, we install Liberty, MDM, and DWC on DR server as well. Database must be installed prior to the HWA implementation on DR server. The installation step for DR setup is same as high availability architecture. Figure 3-10 depicts a typical high availability HWA setup.

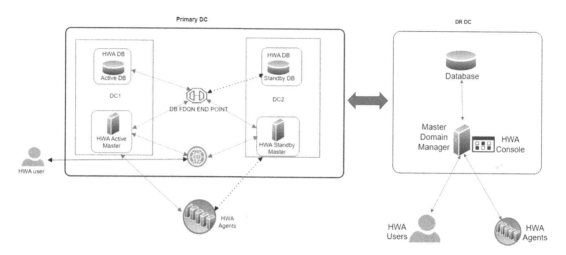

Figure 3-10. *Disaster Recovery Architecture*

The following is the deployment procedure:

- Download the Liberty, HWA, and DWC and db2 package from Flexnet portal.

- Communication between primary and secondary servers of HWA engine and database servers must be enabled as well with DR server.

- Perform the same steps as stand-alone architecture for installing liberty, HWA, and DWC on both primary and DR DC servers.

- In primary data center servers, configure HADR on db2 databases for both primary and secondary servers. Follow the steps detailed in high availability architecture.

- For DR data center, auxiliary standby (DR database) needs to be added in the preceding primary data center DB2-HADR setup. Please follow the following link to configure auxiliary standby database.

  `www.ibm.com/support/pages/steps-adding-new-auxiliary-standby-existing-db2-hadr-pair`

- enAutomaticFailover option has to be set to "Yes" in HWA end using the optman command as the following. By default, this is set to "Yes." Use "optman show af" to show the current value, and use "optman chg af=yes" to set it to yes. The file path is <TWSDATA/mozart>.

  `enAutomaticFailover / af = YES`

- Create the DR server definition as FTA with full status ON as same as backup master server so that real-time sync of job status (symphony) will be updated in DR server as same as primary master server. Refer to Figure 3-5. Symphony file is nothing but production plan file which has the detailed information about the jobs as well as job status information.

    To create the FTA definition with full status "ON," navigate to Design➤ Create Workstation. Refer to Figure 3-11.

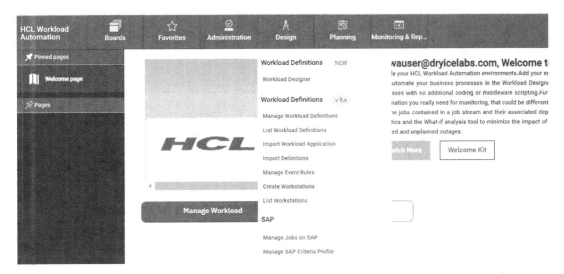

Figure 3-11. *Create Workstation Page*

- This opens the following interface to create the workstation definition; refer to Figure 3-12. Update the DR server details and enable full status on and then select "Save." This will create the new DR server definition in Master Domain Manager.

Figure 3-12. *DR Workstation Properties*

- In case primary data center is down, follow the following steps to enable DR database as primary database and HWA DR server as Primary Master Domain Manager.

Please follow the following steps to enable DR database as primary database server:

```
www.ibm.com/docs/en/db2/11.5?topic=databases-examples-
takeover-in-multiple-hadr-standby-setup
```

Once the DR database is act as primary database server, follow the following steps for HWA DR server to act as a Primary Master Domain Manager.

Log in to the DR server; execute the following commands on DR server to change the DR server role from FTA to Master Domain Manager and also pass on the event processor role to the DR server.

```
switchmgr MASTERDM; DRserver
switcheventprocessor DRSERVER
```

All the agents connected to primary master server will connect to DR server.

- Perform the validation steps as same as stand-alone architecture.

- Once the primary data center is up and running, we have to revert the changes manually for both database and HWA applications.

- Perform the following steps to change the primary data centre primary server as primary database.

  ```
  www.ibm.com/docs/en/db2/11.5?topic=databases-examples-
  takeover-in-multiple-hadr-standby-setup
  ```

- Once the primary data center primary database is acting as primary database, HWA master role has to be changed from DR server to HWA primary data center primary server.

- Follow the following steps to change the master domain manager role

  login to the primary data center primary server, execute the following commands on primary server to change the Master Domain Manager role from DR server, and also pass on the event processor role from the DR server.

  ```
  switchmgr MASTERDM; Primary DC server
  switcheventprocessor DRSERVER
  ```

  All the agents connected to DR master server will connect back to primary data center primary server.

- Perform the validation steps same as stand-alone architecture.

Containerized Deployment

HWA deployment options on containers are available for all platforms. This helps to speed up deployments, upgrade, and replicate the environments when a container fails. Figure 3-13 depicts a containerized high availability HWA setup.

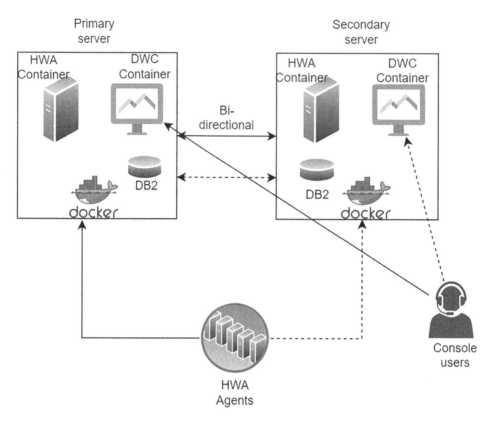

Figure 3-13. *Containerized Deployments*

- Download the HWA and DWC docker images from Flexnet portal.

- Install docker and docker-compose binaries on both the servers.

- Please note that DB2 has to be installed on both the servers, and two database instances have to be created on the database servers. Follow the steps detailed in high availability architecture for deployment.

- Create the database using create_database.sql query. Please note the database to be created using database instance user, so switch to db2inst1 user and run create_database.sql to create the database for HWA and switch to db2inst2 and run create_database.sql for DWC database using following command.

```
db2 -tvf create_database.sql
```

- Load the HWA and DWC images using docker load command as the following:

```
docker load --input workload-automation-console.tar
docker load --input workload-automation-server.tar
```

- Once the images are loaded, modify Docker-compose.yml file and update database, HWA, DWC, and HA information and image names.

- Please find the following attached docker-compose.yml file for both MDM and DWC.

```
MDM yml file:


###########################################################
##########
# Licensed Materials - Property of HCL*
# (c) Copyright HCL Technologies Ltd. 2018-2020. All
rights reserved.
#
# * Trademark of HCL Technologies Limited
###########################################################
##########
version: '3.5'
services:
  wa-server:
   environment:
      - LICENSE=notaccept # Use ACCEPT to accept the
license agreement
      - PUBLIC_PORT=31116 # The public https port to
reach server in this container. Must be the container
external port you choose to map to the container internal
https port.
        - WA_PASSWORD=<password> # The password of the
"wauser" user used to connect to the server (server will
be installed with this password)
```

```
      - AGT_NAME=WA_AGT # The name to assign to the dynamic
agent of the server (optional)
      - DATE_FORMAT=MM/DD/YYYY # The date format assigned
to the plan (optional)
      - CREATE_PLAN=true # If true, it executes an automatic
JnextPlan at container deploy time (optional)
      - COMPANY_NAME=my-company # The company name
(optional)
      - DB_TYPE=DB2 # The remote database server type you
choose. (DB2 | ORACLE | MSSQL | IDS)
      - DB_HOSTNAME=<dbhostname> # The host name or IP
address of database server
      - DB_PORT=50000 # The port of the database server
      - DB_NAME=TWS # For DB2/IDS/MSSQL is the name of
server database. For ORACLE is the Oracle Service Name
      - DB_USER=db2inst1 # Database user that accesses the
server tables on the database server
      - DB_PASSWORD=<password> #  The password of the
database user that accesses the server tables on the
database server
      - DB_ADMIN_USER=db2inst1 # Database user administartor
that accesses the server tables on the database server
      - DB_ADMIN_PASSWORD=<password> #  The password of
the database user administrator that accesses the server
tables on the database server
      - LANG=en # The language of the server (en | de | es
| fr | it | ja | ko | pt_BR | ru | zh_CN | zh_TW)
      - SERVERHOSTNAME=wa-server #  The host name on which
the server is contacted by internal dynamic agents (do
not touch)
      - SERVERPORT=31116 #  The port on which the server is
contacted by internal dynamic agents (do not touch)
      # - LICENSE_SERVER_URL=<license server URL> # URL
of the license server which processes license usage
information
```

```
    # - LICENSE_SERVER_ID=<license server id> # ID of the
license server which processes license usage information
    # - PUBLIC_HOSTNAME=<pubhostname> # The fully qualified
host name or IP address on which the server is contacted
by external dynamic agents
    # - EVENTPROCESSOR_HOSTNAME=<evtprocpuhostname> # The
fully qualified host name or IP address on which the event
processor is contacted by external dynamic agents
    # - TZ=Europe/Rome # The TZ operating system
environment variable
    # - START_OF_DAY=0700 # The start time of the plan
processing day in 24 hour format: "hhmm" (optional)
    # - TIMEZONE=Europe/Rome # The Time Zone for the start
time of the plan processing day (optional - used only if
START_OF_DAY is defined)
    # - DB_SSL_CONNECTION=true # If true, enables SSL
connection to remote database server. Customized trust JKS
file is needed. Valid only for DB2 (optional)
    # - DB_TS_NAME= # The TWS name space's name (optional)
    # - DB_TS_PATH= # The TWS name space path's name.
Valid only for DB2/IDS/MSSQL (optional)
    # - DB_LOG_TS_NAME= # The LOG name space's name
(optional)
    # - DB_LOG_TS_PATH= # The LOG name space path's name.
Valid only for DB2/IDS/MSSQL (optional)
    # - DB_PLAN_TS_NAME= # The PLAN name space's name
(optional)
    # - DB_PLAN_TS_PATH= # The PLAN name space path's
name. Valid only for DB2/IDS/MSSQL (optional)
    # - DB_TEMP_TS_NAME= # The TEMP name space's name.
Valid only for ORACLE (optional)
    # - DB_SBSPACE= The SBSPACE space's name. Valid only
for IDS (optional)
```

```
    # - DB_ENABLE_PARTITIONING_OPTION=true # If true, the
Oracle Partitioning feature is enabled. Valid only for
Oracle, it is ignored by other databases. The default
value is true
    # - DB_SKIP_CHECK=true # If you want to skip db check
it must be set TRUE. Valid only for Oracle.
    # - LICENSE_SERVER_ID= # License Server ID valid only
for HCL distribution.
    image: hcl-workload-automation-server:10.1.0.00
    ports: # port mapping "external:internal". Internal
ports are predefined.
      - "31116:31116" #HTTPS SERVER
      - "31111:31111" #NETMAN
      - "31131:31131" #EIF
    # - "35131:35131" #EIF SSL
    container_name: "wa-server"
    hostname: "wa-server"
    networks:
      - wa-net
    volumes:
      - wa-server-data:/home/wauser
    logging:
      driver: "json-file"
      options:
        max-size: "10m"
        max-file: "5"
volumes:
  wa-server-data:
networks:
  wa-net:
    name: wa-net
```

DWC yml file:

```
    ####################################################
###############
# Licensed Materials - Property of HCL*
```

```
# (c) Copyright HCL Technologies Ltd. 2018-2020.
All rights reserved.
#
# * Trademark of HCL Technologies Limited
#######################################################
#############
version: '3.5'
services:
  wa-console:
    environment:
      - LICENSE=notaccept # Use ACCEPT to accept the
license agreement
      - WA_PASSWORD=<password> # The password of the
"wauser" user used to connect to the console (console will
be installed with this password)
      - DB_TYPE=DB2 # The remote database server type you
choose. (DERBY | DB2 | ORACLE | MSSQL | IDS)
      - DB_HOSTNAME=<dbhostname> # The host name or IP
address of database server. Valid only for DB2/IDS/
MSSQL/ORACLE
      - DB_PORT=50000 # The port of the database server.
Valid only for DB2/IDS/MSSQL/ORACLE
      - DB_NAME=DWC # For DB2/IDS/MSSQL is the name of
console database. For ORACLE is the Oracle Service Name.
Valid only for DB2/IDS/MSSQL/ORACLE
      - DB_USER=db2inst1 # Database user that accesses the
console tables on the database server. Valid only for DB2/
IDS/MSSQL/ORACLE
      - DB_PASSWORD=<password> #  The password of the
database user that accesses the console tables on the
database server. Valid only for DB2/IDS/MSSQL/ORACLE
      - DB_ADMIN_USER=db2inst1 # Database user administartor
that accesses the console tables on the database server.
Valid only for DB2/IDS/MSSQL/ORACLE
```

```
    - DB_ADMIN_PASSWORD=<password> #  The password of
the database user administrator that accesses the console
tables on the database server. Valid only for DB2/IDS/
MSSQL/ORACLE
    - LANG=en # The language of the console container
(en | de | es | fr | it | ja | ko | pt_BR | ru | zh_CN
| zh_TW)
    # - ENGINE_HOSTNAME= # The host name or IP address of
the engine (optional)
    # - ENGINE_PORT= # The host name or IP address of the
engine (optional, default 31116)
    # - ENGINE_USER= # The user of the engine (optional,
default wauser)
    # - TZ=Europe/Rome # The TZ operating system
environment variable
    # - DB_SSL_CONNECTION=true # If true, enables SSL
connection to remote database server. Customized trust JKS
file is needed. Valid only for DB2 (optional)
    # - DB_TS_NAME= # The TWS name space's name. Valid
only for DB2/IDS/MSSQL/ORACLE (optional)
    # - DB_TS_PATH= # The TWS name space path's name.
Valid only for DB2/IDS/MSSQL (optional)
    # - DB_SBSPACE= The SBSPACE space's name. Valid only
for IDS (optional)
    # - DB_ENABLE_PARTITIONING_OPTION=true # If true, the
Oracle Partitioning feature is enabled. Valid only for
Oracle, it is ignored by other databases. The default
value is true
    # - DB_SKIP_CHECK=true # If you want to skip db check
it must be set TRUE. Valid only for Oracle.
    image: hcl-workload-automation-console:10.1.0.00
    ports: # port mapping "external:internal". Internal
ports are predefined.
     - "9443:9443" #HTTPS CONSOLE
    container_name: "wa-console"
```

```
      hostname: "wa-console"
      networks:
       - wa-net
      volumes:
       - wa-console-data:/home/wauser
      logging:
        driver: "json-file"
        options:
          max-size: "10m"
          max-file: "5"
  volumes:
    wa-console-data:
  networks:
    wa-net:
      name: wa-net
```

- Execute "docker-compose up" command to create the containers for HWA and DWC. Perform the same steps on secondary server.

- Db2 HADR configuration has to be completed on both the servers for the db2 instances.

- Validate both the consoles are working fine as same as stand-alone architecture.

HWA Deployment on Cloud

To ensure a fast and responsive experience when using HCL workload automation, you can deploy HCL workload automation on a cloud infrastructure. A cloud deployment ensures access anytime anywhere and is a fast and efficient way to get up and running quickly. It also simplifies maintenance, lowers costs, provides rapid upscale and downscale, and minimizes IT requirements and physical on-premises data storage.

As more and more organizations move their critical workloads to the cloud, there is an increasing demand for solutions and services that help them easily migrate and manage their cloud environment.

Please refer the following link for more details about this deployment.

GitHub (`https://github.com/WorkloadAutomation/hcl-workload-automation-chart`)

Summary

In this chapter, we covered various deployment options available to us for deploying HWA in our environment. We covered the following in detail. The readers will be able to choose the best deployment option and deploy HWA in their environment.

- Various deployment options available for HWA

- Recommendations to choose a suitable architecture

- Detailed steps to implement different HWA architectures

- DB2 implementation with HADR setup

CHAPTER 4

Workload Design and Monitoring Using DWC

This chapter explains in detail about the HCL workload automation front-end console DWC. We will cover in detail how to design, monitor, and administer jobs and job streams using the graphical user interface.

Note At the end of this chapter, readers are expected to have good understanding on designing, monitoring, and managing their workflows using HWA.

Designing and Monitoring Workload Objects Using HWA Dynamic Workload Console

HCL workload automation (HWA) is a batch automation and scheduling tool which is used to run application jobs like SAP, Native, Azure, Kubernetes, database, REST APIs, etc., apart from the normal command jobs, wrappers, file transfer jobs, etc.

HWA provides a graphical user interface called Dynamic Workload Console (DWC) along with command-line interface (CLI). This console facilitates creating, designing, and managing the workload objects. Dynamic Workload Console can be accessed using a web browser.

Format of the URL for accessing the console in latest version is as the following:

```
https://<IP address or hostname of the DWC host server>:9443/console/
login.jsp
```

Login Page of Dynamic Workload Console is shown in Figure 4-1.

49

© Navin Sabharwal and Subramani Kasiviswanathan 2023
N. Sabharwal and S. Kasiviswanathan, *Workload Automation Using HWA*,
https://doi.org/10.1007/978-1-4842-8885-6_4

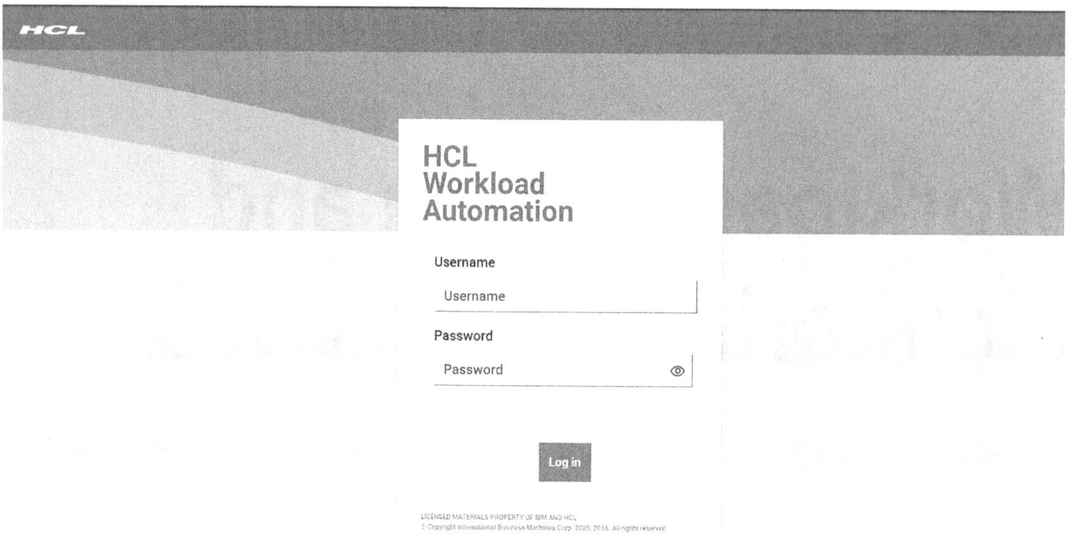

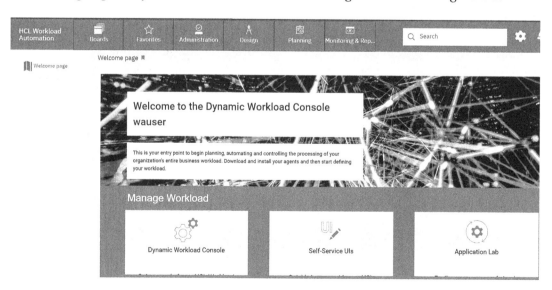

**HCL
Workload
Automation**

Username

> Username

Password

> Password

Log in

Figure 4-1. *HWA Login Page*

Note Username and password must be provided to log in to the console. Username must be provided with necessary roles by the administrator to access the console.

Landing Page of Dynamic Workload Console after login is shown in Figure 4-2:

Figure 4-2. *Landing Page of Dynamic Workload Console*

There are four tabs in the console: Administration, Design, Planning, and Monitoring and Reporting.

Please find the following detailed explanation of the preceding tabs.

Administration tab:

Administration tab comprises of following options:

Manage engines: For creating and managing the HWA engine definitions.

Managing workload security: For managing access lists, folders, security domains, and roles. Through access control list, we can manage access for users, groups, and security domains.

Auditing preferences: HWA operations can be audited using this option.

Manage settings: For exporting and importing user settings in Dynamic Workload Console in .xml format. User settings such as user preferences, saved tasks, and engine connections are stored in the settings repository.

Manage roles: For customizing the roles in the console.

Design tab:

Design tab comprises of following options:

Manage workload Definitions: For creating, editing, and deleting the Scheduler objects in database.

List workload definitions: To list job and job stream definitions.

Import workload application: For importing workload application template that contains a workload process that is used to replicate all the required objects in the new environment.

Import definitions: For importing Cron/Windows Task scheduler jobs into HWA scheduling distributed environment. DWC reads the imported definitions and converts them into HWA scheduler objects.

Create event rules: We can create and run an event monitoring task by specifying a filter requirement for getting the information about required objects.

Manage event rules: Option to view the configured events and to manage them.

Create workstations: Option to create workstation definitions.

List workstations: To view the configured workstations and to edit and to manage them.

Manage jobs on SAP: To view the jobs on SAP systems which are integrated with HWA.

Planning tab:

Plan is considered as the actual execution of the jobs created on the database. Plan consists of jobs, job stream, etc. Everyday plan is created in HWA based on the requirements.

Manage available plans: To retrieve or display the already available plans (archived, trial, and forecast).

Create trial plan:

A trial plan comprises of potential problems and offers the opportunity to avoid problems before they affect your business systems. You should create a trial plan before everyday plan extends. This trial acts as the early warning message for the actual plan. A trial plan is a created for longer period of time to validate how the production will look like. For example, if you generate a production plan that covers 2 days, but you want to know what the plan would be if it covered 3 days, you could generate a trial plan.

Create forecast plan:

Forecast plan is created for the specific time interval to validate how the actual production plan will look like. For example, if you generate a production plan that covers 2 days and you want to know what the plan would be for the next week, you can generate a forecast plan. Future job execution plan can be created using forecast plan options; this option is used during patching, upgrade, etc.

View preproduction plan:

> Use this option to access the preproduction plan. Preproduction plan is used to identify in advance the job stream instances, and the job stream dependencies involved in a specified time.

Workload submission:

In workload submission, you can submit the jobs and job streams on ad hoc basis.

> Job: Job is a unit of script/command which runs on the target system.

> Job stream: Group of jobs mapped to execute on certain timeframe. Based on the user requirement, we can schedule the jobs in a job stream.

> Calendar: To run the jobs on required frequency, we can use calendar option. We can run the jobs on daily, monthly, or weekly schedules.

Submit ad hoc jobs:

> Submit ad hoc job option is used when we have a request to test the specific script in a server and the job is not predefined on the agent. In this kind of scenarios, we can use submit ad hoc job option to execute those scripts and share the results to the user.

Submit predefined jobs:

> This option is used whenever there is a request to run additional execution of job in plan which is already created in database. We need to submit this job using submit predefined job option.

Submit predefined job streams:

> This option is same as submitting a predefined job; however, in this case we are submitting the predefined job stream. We can use Submit predefined job stream option.

Monitoring and reporting tab:

> This section explains the details about monitoring of HWA objects and reporting. Please see the following high-level explanations:

Monitor workload:

To manage the workload objects in current plan and to view the objects in archived plans.

Show plan view:

This option displays a concise view of the plan showing only the job streams and dependencies flow between them.

Monitor event rules:

Status of events can be monitored. Event rule is nothing but a set of actions that are to run when a specific event condition is met. Event rules are created to trigger actions like sending mail, creating the tickets in snow for job failures, submitting a job stream/job, etc. The conditions are generally file dependency, job failures, job long running, etc.

Monitor operator messages:

One of the available event rule actions is to write a message for operators in the HWA database when the rule conditions are met.

This part provides the possibility to see operator messages written by event rule instances.

Monitor triggered actions:

Once the Event rule is satisfied, the HWA event processor will take the action described by the event rule action definition. This could be sending an email, submitting a job stream, or executing a command. Most of the HWA event actions are pretty straightforward to use. To view the triggered actions configured in the events when the event conditions have been satisfied.

Manage predefined reports:

To run the predefined reports available in Dynamic Workload Console.

Manage personalized reports:

To create the personalized reports.

Personalized reports:

Run the personalized reports created with Business Intelligent Report Tool (BIRT) based on the requirement.

Alter user password in plan:

For Windows jobs, we use Windows users to run the actual scripts. In case of any change in password for the Windows users, we must update them in HWA plan. If we failed to update the password in HWA, it will lead to job failure. To update the password in HWA plan, we can use this option.

Business Requirement

To understand the usage of workload automation, let us consider a real-time use case where a customer has the following requirement to create and schedule a business process workflow to run on a specific calendar.

Use case 1:

Table 4-1 lists the details of the requirement.

Table 4-1. *Business requirement*

Workstation	Folder	Job name	Stream Logon\User ID	Script \ DoCommand	Job Stream	Frequency	Dependency	Start time
DA	HW	HWA_JOB	wauser	/HWA/WA/script.sh	HWA_JS	DAILY	NA	7:00 AM
DA	HW	HWA_JOB1	wauser	hostname	HWA_JS	DAILY	Follows HWA_JOB	NA
DA	HW	HWA_JOB2	wauser	sleep 40	HWA_JS	DAILY	Follows HWA_JOB1	NA
DA	HW	HWA_JOB3	wauser	/HWA/WA/testing.sh	HWA_HJS	WEEKLY(FRIDAY)	Follows HWA_JS#HWA_JOB2	NA
DA	HW	HWA_JOB4	wauser	pwd	HWA_HJS	WEEKLY(FRIDAY)	Follows HWA_JOB3	NA

From the preceding requirement, we are expected to create five jobs and two streams with internal and external dependency on daily and weekly calendars.

Internal dependencies are limited to job dependencies in the same job stream, while External dependencies are job dependencies from a different job stream.

Please see the following procedure to create the jobs; to create the job in Dynamic Workload Console, follow the following steps:

navigate to Design ➤ Manage Workload Definition ➤ Select the engine ➤ Create New ➤ Job definition.

Manage Workload Definitions provides us an interface to design workload.

Please see Figure 4-3 for reference.

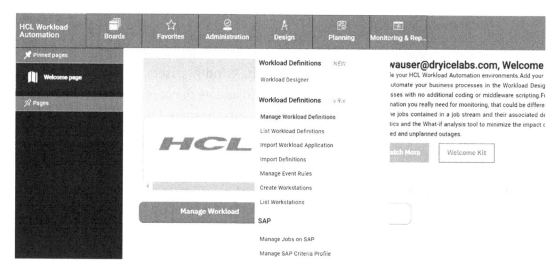

Figure 4-3. *Manage Workload Definition Page in DWC.*

In the manage workload definitions page, select the Master Domain Manager engine name to which we want to connect as shown in Figure 4-4.

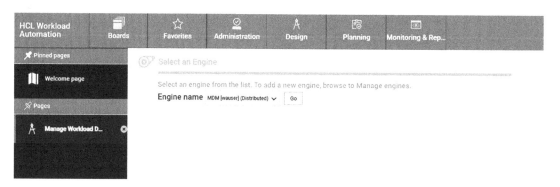

Figure 4-4. *Engine Selection*

The aforementioned action opens the following interface which is a workload designer where you can design your job.

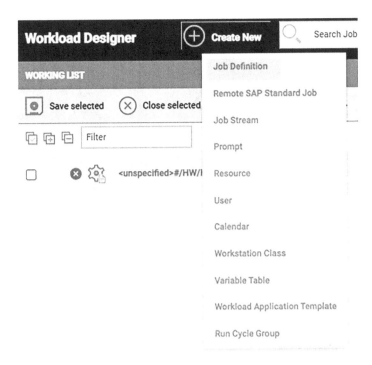

Figure 4-5. *Workload Designer Page*

- Select the type of job that needs to be created as shown in Figure 4-6; in our case, it is a Unix job and workstation name is "DA." Click "OK."

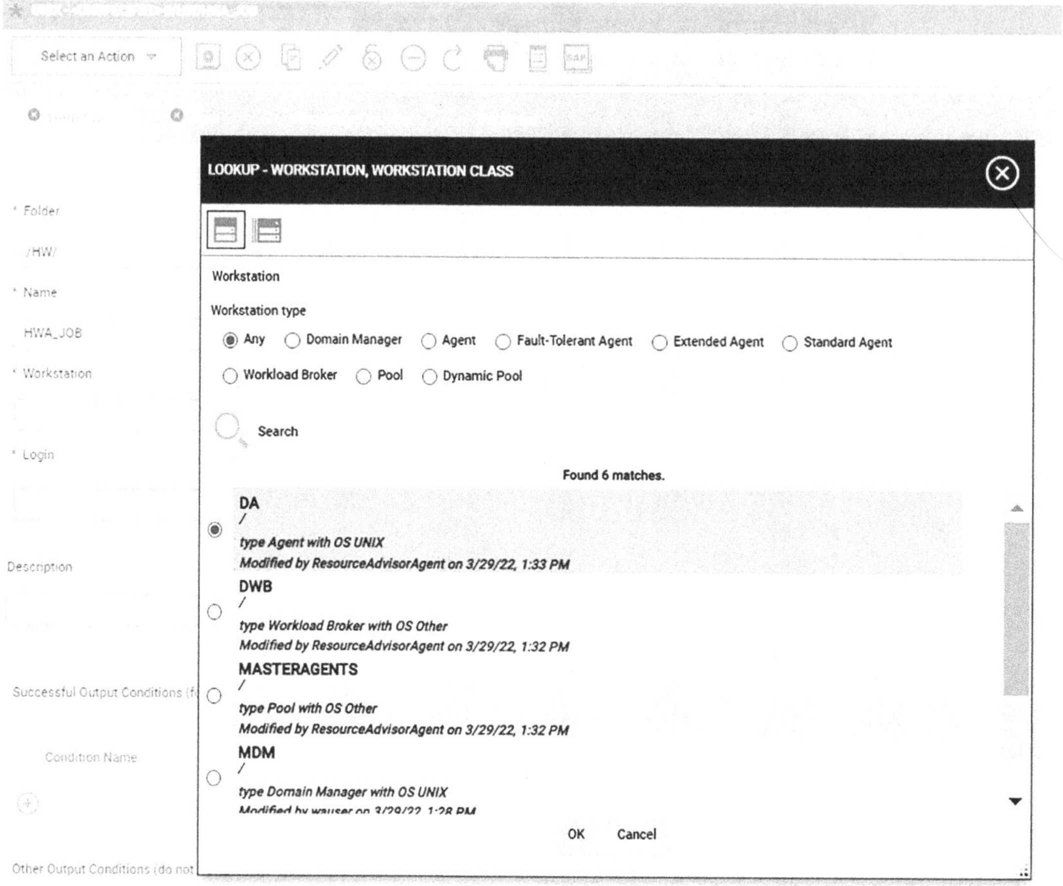

Figure 4-6. *Agent and Job-Type Selection Page*

The following interface opens where you can define the job properties. Provide the job parameters as per requirements as shown in Figure 4-7.

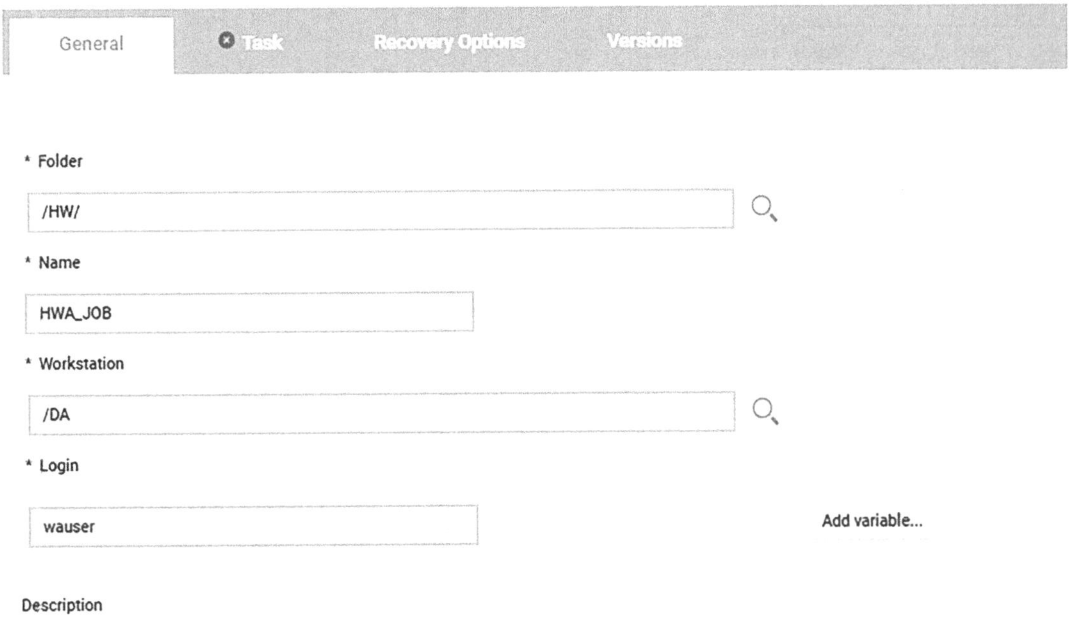

Figure 4-7. *Job Definition Properties*

- Update the script path on task properties. This is the executable that needs to run on an agent machine as per the schedule definitions as shown in Figure 4-8. Alternatively, this can also be a command that needs to be run on an agent machine like "ls -lrt" or "pwd."

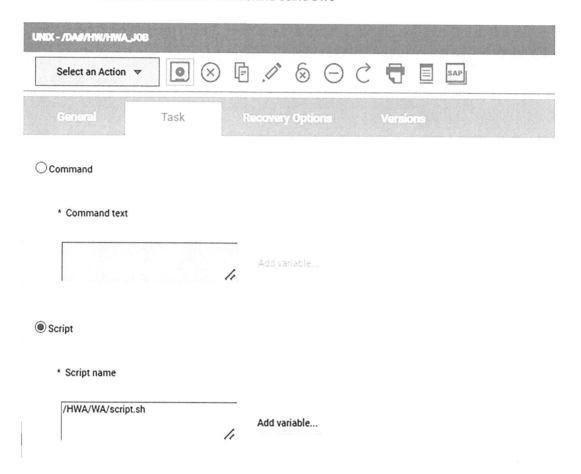

Figure 4-8. *Job Definition Task View*

- Click save to create the job and follow the same steps for other four jobs.

- Once remaining four jobs are created, create the job stream.

- To create the job stream, navigate to Manage Workload Definition ➤ Select Engine ➤ select create new ➤ Job stream option. Please see Figure 4-9 for job stream creation.

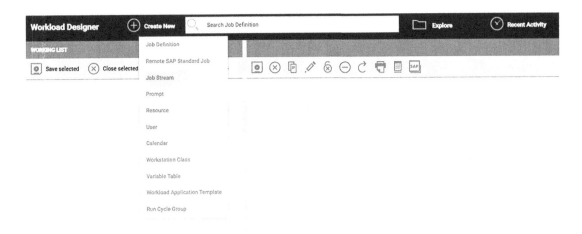

Figure 4-9. *Job Stream Selection View*

- Select job stream option as shown in Figure 4-9, to create the job stream as per the request.

- Select workstation name as "DA" as shown in Figure 4-10 and click "OK."

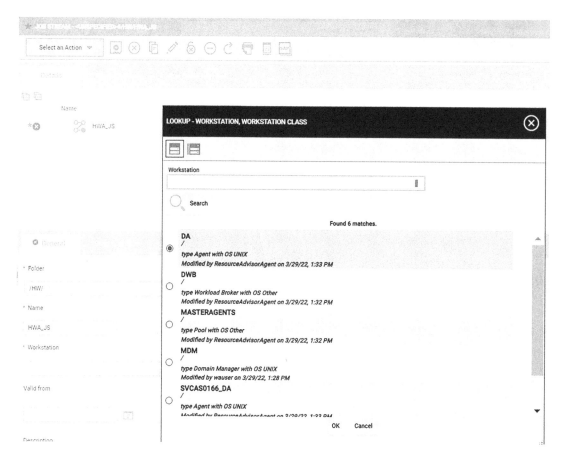

Figure 4-10. *Workstation Selection*

- Figure 4-11 shows the properties of the job stream:

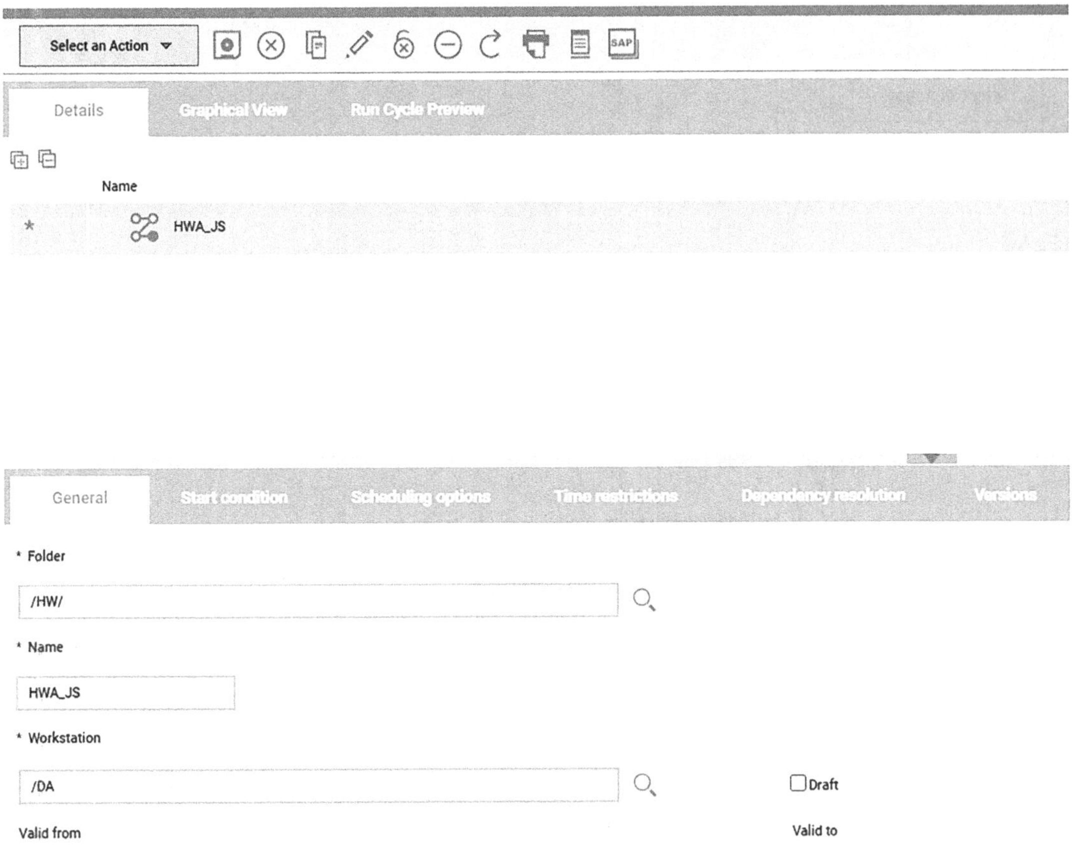

Figure 4-11. *Job Stream Properties*

- Once the stream is defined, add the run cycle as shown in Figure 4-12.

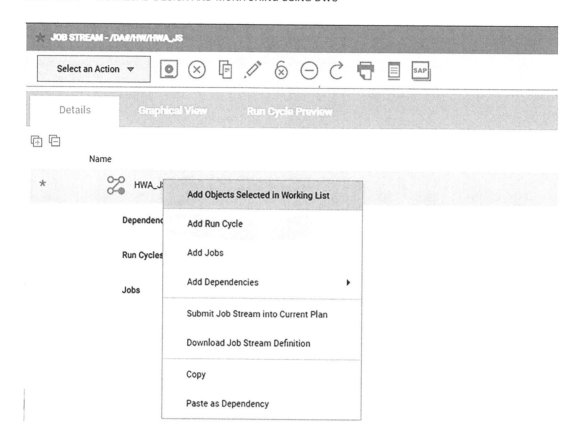

Figure 4-12. *Run Cycle Selection View*

Run Cycle:

> Run cycles are added for the frequency of the jobs to be executed.
> In run cycles, we have the option to select daily, weekly,
> monthly, and yearly run cycles in HWA. In our case we have two
> frequencies to be selected, i.e., daily and weekly. Daily run cycle
> is to be configured for HWA_JS job stream and weekly run cycle to
> be configured for HWA_HJS.

- Select the daily calendar for the run cycle as per the request for the job stream "HWA_JS" as shown in Figure 4-13.

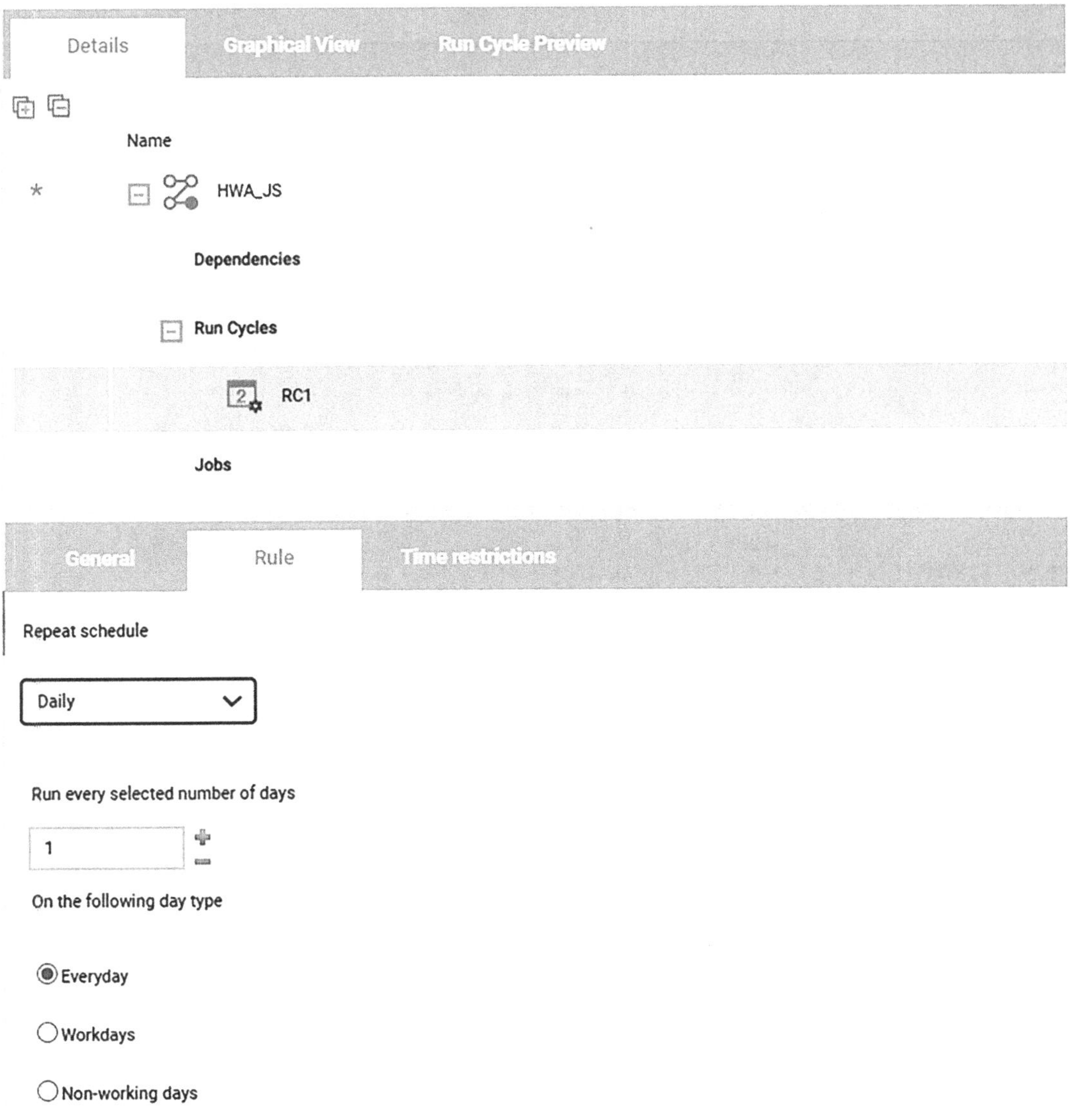

| Details | Graphical View | Run Cycle Preview |

Name

* ☐ HWA_JS

Dependencies

☐ Run Cycles

[2] RC1

Jobs

| General | Rule | Time restrictions |

Repeat schedule

| Daily ∨ |

Run every selected number of days

| 1 | ➕➖ |

On the following day type

◉ Everyday

○ Workdays

○ Non-working days

Figure 4-13. *Run Cycle Rule Selection View*

65

We have successfully created the run cycle and the run cycle preview as shown in Figure 4-14. We can see that all days are selected for executing the job as defined in the requirements.

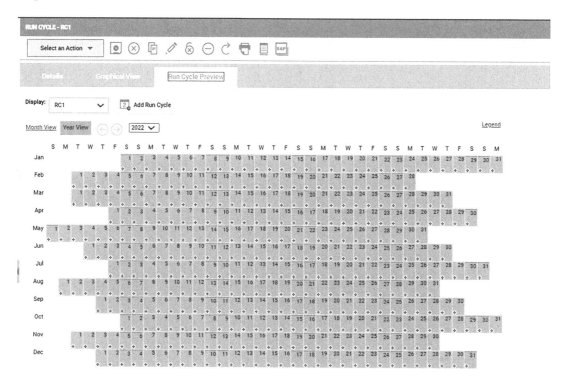

Figure 4-14. *Run cycle Preview*

- We have successfully created the job stream by adding a run cycle. Now we need to add the required jobs into the job stream as shown in Figure 4-15 as per the request.

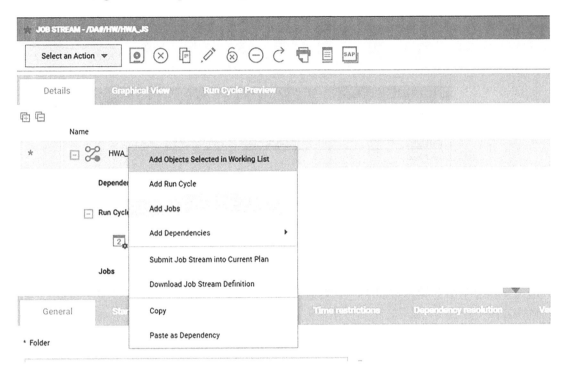

Figure 4-15. *Job Selection Option*

- The following interface opens where you can add the jobs into the job stream. Search the jobs that are required to be added as shown in Figure 4-16.

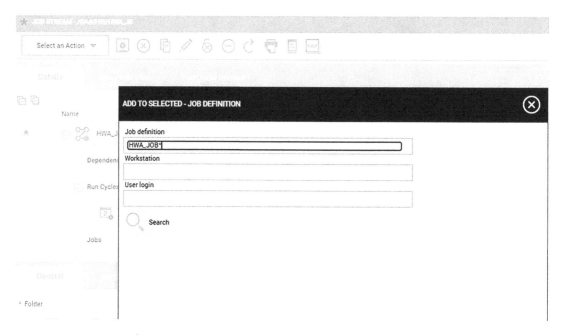

Figure 4-16. *Addition of Jobs into Job Stream View*

- Select the jobs that are to be added in the job stream in Figure 4-17 and click Add.

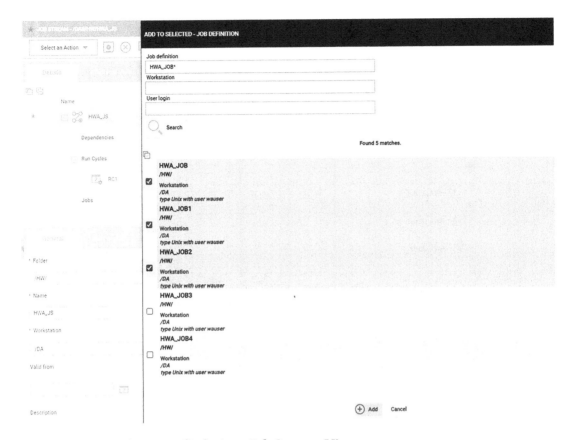

Figure 4-17. *Selection of Jobs into Job Stream View*

- We have successfully added the jobs HWA_JOB, HWA_JOB1, and HWA_JOB2 on the job stream HWA_JS.

- As defined in the use case, provide the start time condition as 7 AM for HWA_JOB shown in Figure 4-18.

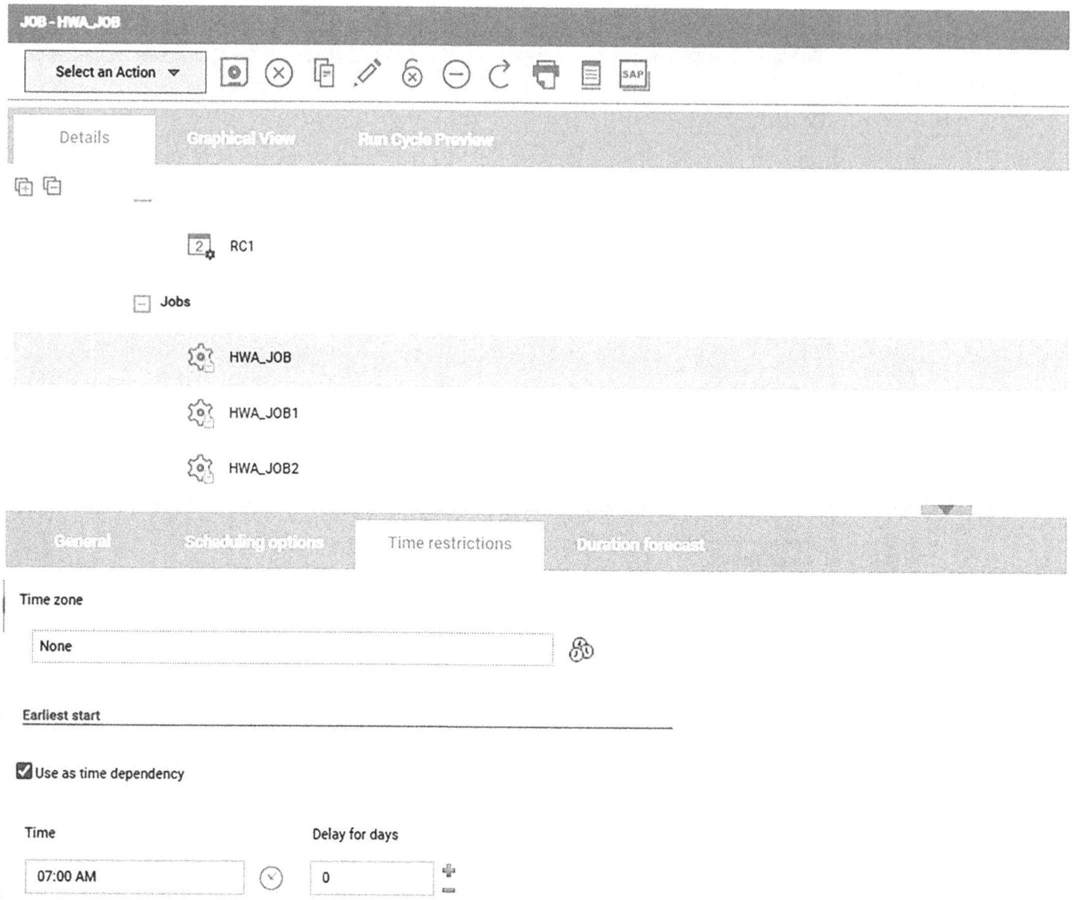

Figure 4-18. *Start Time Condition View of Job*

To execute the job automatically on specific time, we can use time restriction option.

- Add the dependencies as per the use case. For HWA_JOB1, it has a dependency of HWA_JOB, and HWA_JOB is scheduled on same stream. Since the HWA_JOB job is scheduled on same job stream, we can right-click HWA_JOB1 and navigate to Add dependencies ➤ job in the same job stream option and add the HWA_JOB as the dependency. Please refer to Figure 4-19 for reference.

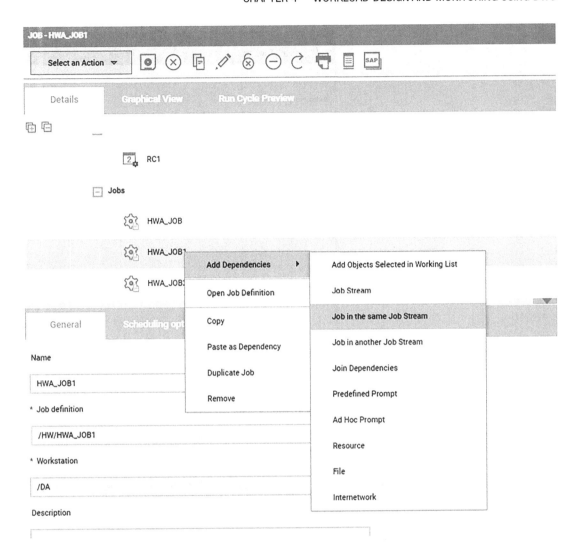

Figure 4-19. *Dependency Options*

- This opens the following interface where you can add the predecessor job for HWA_JOB1 into the same job stream. Search the predecessor job HWA_JOB and add to the job stream as shown in Figure 4-20.

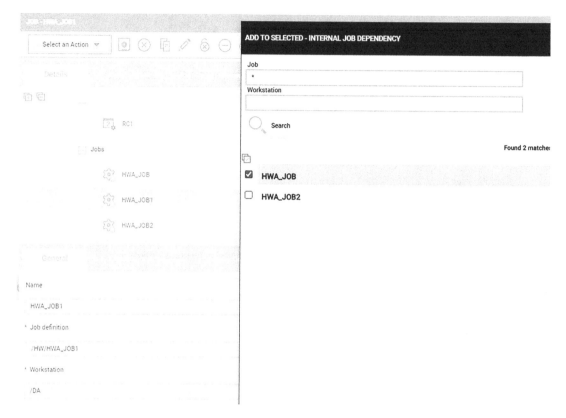

Figure 4-20. *Adding Dependency Jobs*

- Follow the same steps for HWA_JOB2 job and add the predecessor job as HWA_JOB1 for HWA_JOB2 job.

- We have successfully created "HWA_JS" job stream and graphical view of the job stream HWA_JS shown in Figure 4-21.

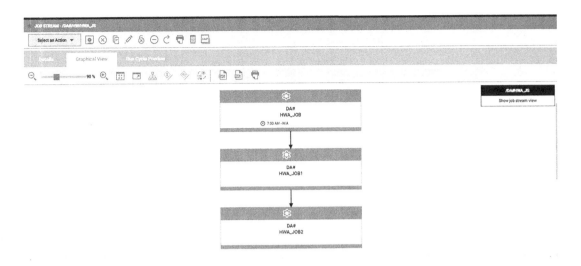

Figure 4-21. *Job Stream Graphical View*

- From Figure 4-21, we could see that HWA_JOB is starting at 7 AM and its successor jobs are HWA_JOB1 and HWA_JOB2.

- We can create a second job stream HWA_HJS as per requirement; please see the following steps for the same.

- Figure 4-22 shows the properties of a job stream provided by the use case.

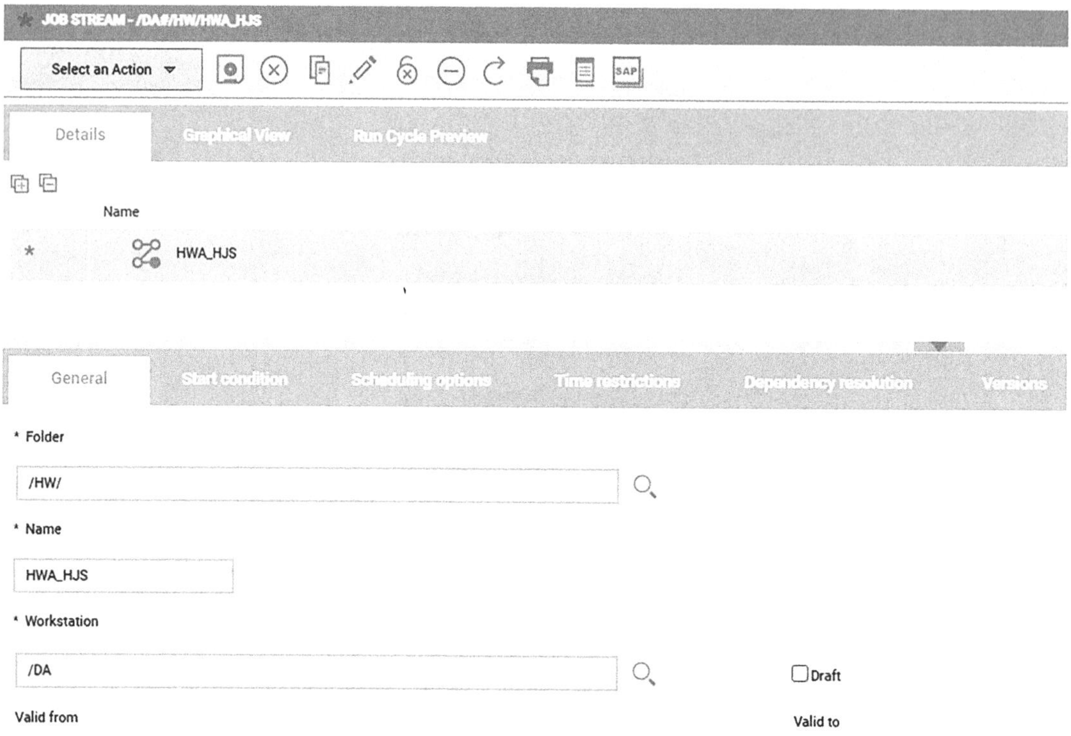

Figure 4-22. *Job Stream Properties*

- Once the stream is defined, add the run cycle as in Figure 4-23.

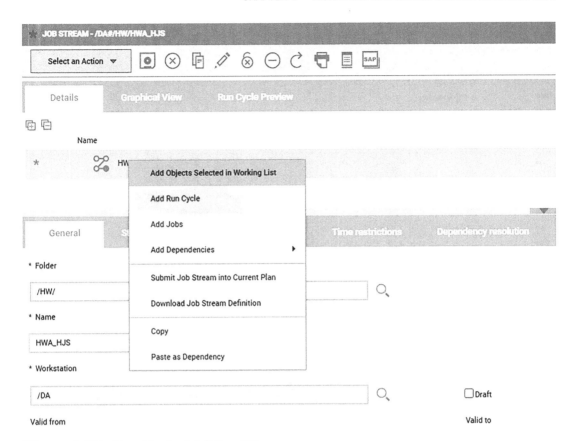

Figure 4-23. *Run Cycle Addition View*

- Select the weekly calendar for the run cycle for the job stream "HWA_
  HJS" and the run cycle preview as shown Figure 4-24. We can see that
  only Friday is selected.

Figure 4-24. *Run Cycle Preview*

- Select the jobs that are to be added in the job stream HWA_HJS in our case search for HWA_JOB3 and HWA_JOB4 and click add. Please see Figure 4-25 for reference.

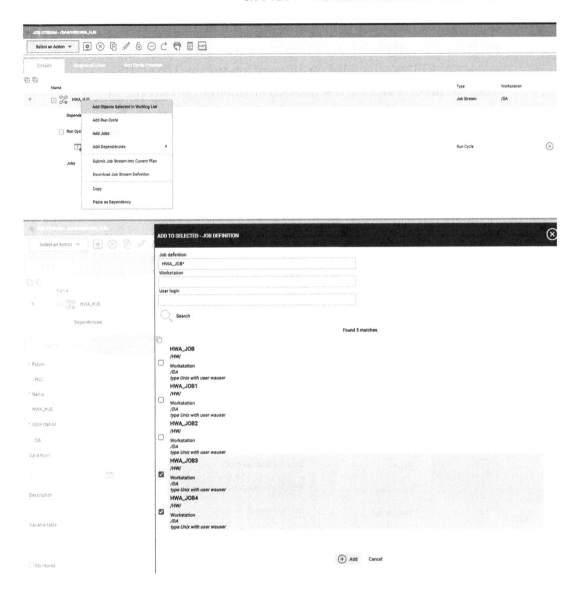

Figure 4-25. *Addition of Jobs into Job Stream*

- Once the jobs are added to the job stream, add the dependencies. For HWA_JOB3, there is a predecessor dependency of HWA_JOB2, and HWA_JOB2 is scheduled on different job stream (HWA_HJS). Since the predecessor job is scheduled on different job stream, we can right-click HWA_JOB3 and navigate to Add dependencies ➤ job in the another job stream option and add the HWA_JOB2 as the predecessor. Please refer to Figures 4-26 and 4-27 for reference.

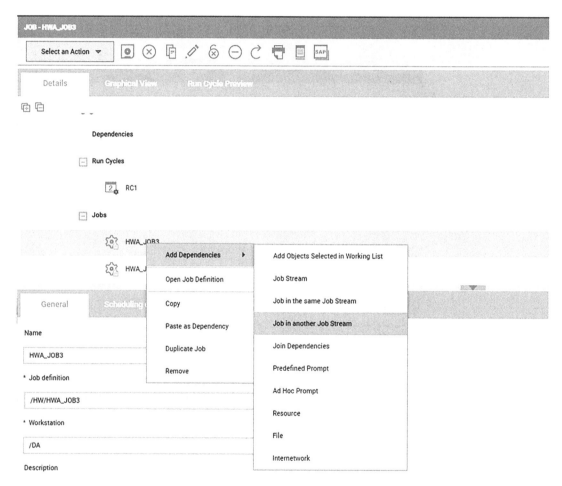

Figure 4-26. *Dependency Selection*

- The above interface opens where you can add external dependency jobs into the same job stream. Search the jobs that are required to be added; in our case, HWA_JOB2 has to be added as a predecessor for HWA_JOB3. Please refer to Figure 4-27.

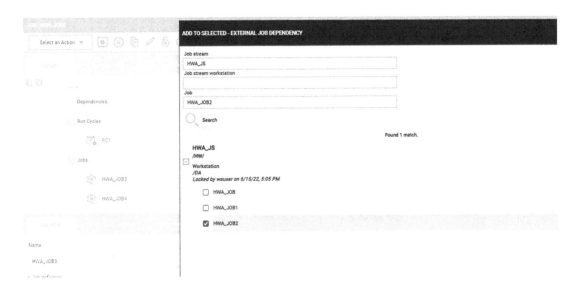

Figure 4-27. *External Job Dependency Selection*

- The HWA_JOB4 has a predecessor dependency of HWA_JOB3, and HWA_JOB3 is scheduled on same stream. Since the predecessor job is scheduled on same job stream, we can right-click on HWA_JOB4 and navigate to Add dependencies ➤ job in the same job stream option and add the HWA_JOB3 as the predecessor. Please see Figures 4-28 and 4-29 for reference.

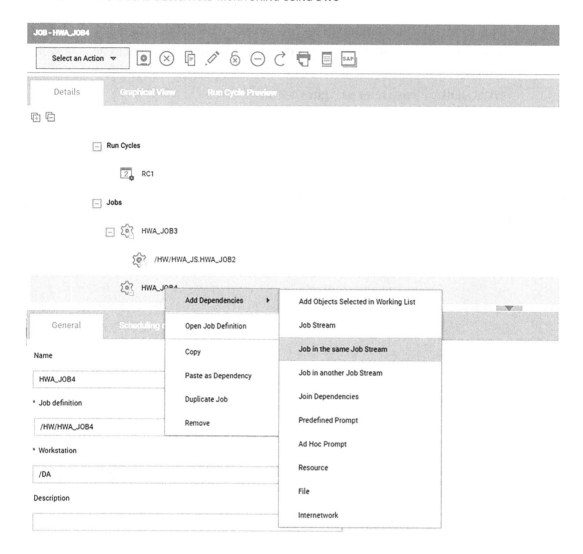

Figure 4-28. *Internal Job Dependency Selection*

- The above interface opens where you can add internal dependency jobs into the same job stream. Search the jobs that are required to be added; in our case, HWA_JOB3 has to be added as a predecessor for HWA_JOB4. Please refer to Figure 4-29.

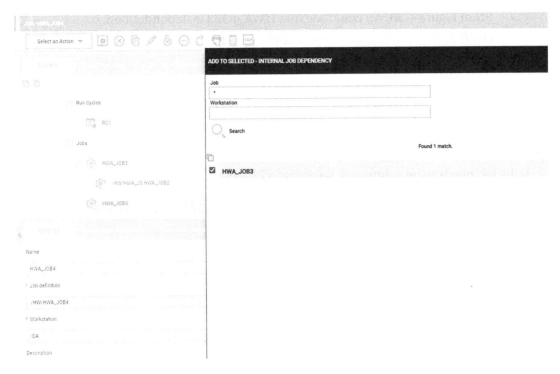

Figure 4-29. *Job Dependency Inside Job Stream*

- We have successfully created "HWA_HJS" job stream and graphical view of the job stream HWA_HJS shown in Figure 4-30.

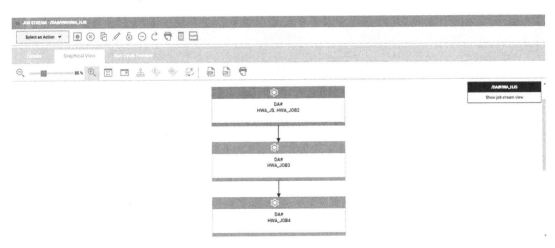

Figure 4-30. *Job Stream Graphical View*

- From Figure 4-30, we could see that HWA_JOB3 is dependent on HWA_JOB2 from HWA_JS stream and HWA_JOB4 is dependent on HWA_JOB3 job from the same job stream.

We have successfully created the jobs and job streams as per the request.

We can submit the predefined jobs and job streams in HWA using the following procedure. This procedure helps to submit the jobs and job steams which are created in HWA using Dynamic Workload Console.

If the job needs to be submitted, navigate to Planning ➤ Submit Predefined job as the following:

Figure 4-31. *Submit Job Option*

- Predefined job is nothing but the job which is already created based in the database.

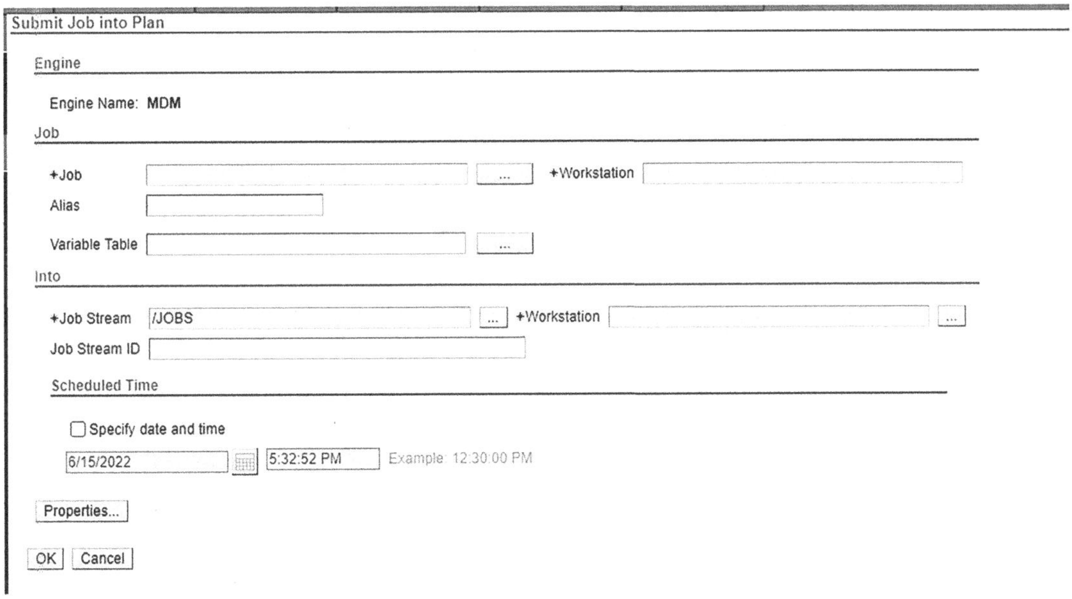

Figure 4-32.   *Submit Job Properties*

- The following interface opens where you can search for the job to be submitted. In our case we are submitting the job HWA_JOB. Please see Figure 4-33 for reference. Select the job and click "ok."

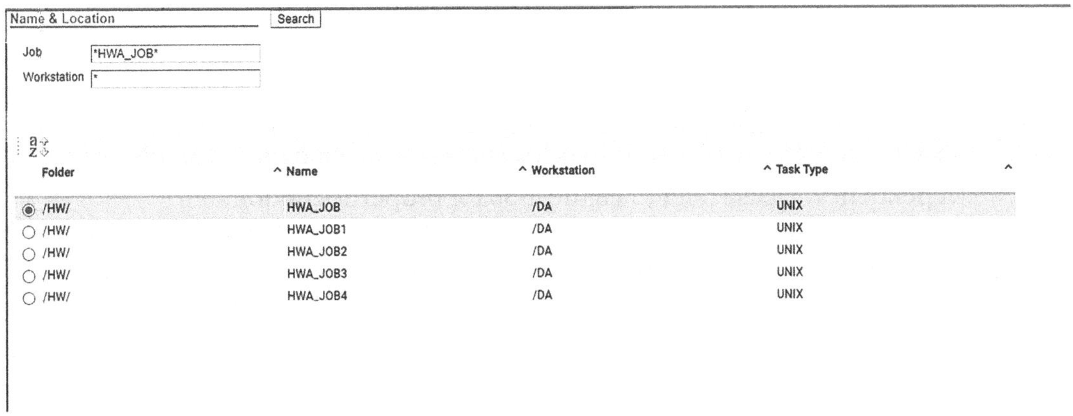

Figure 4-33.   *Selection of Job*

- By default, all the submitted jobs run on "JOBS" job stream; in case, we need to submit the job into different job stream, we can mention the job stream under the job stream ID option. Please see Figure 4-34 for reference.

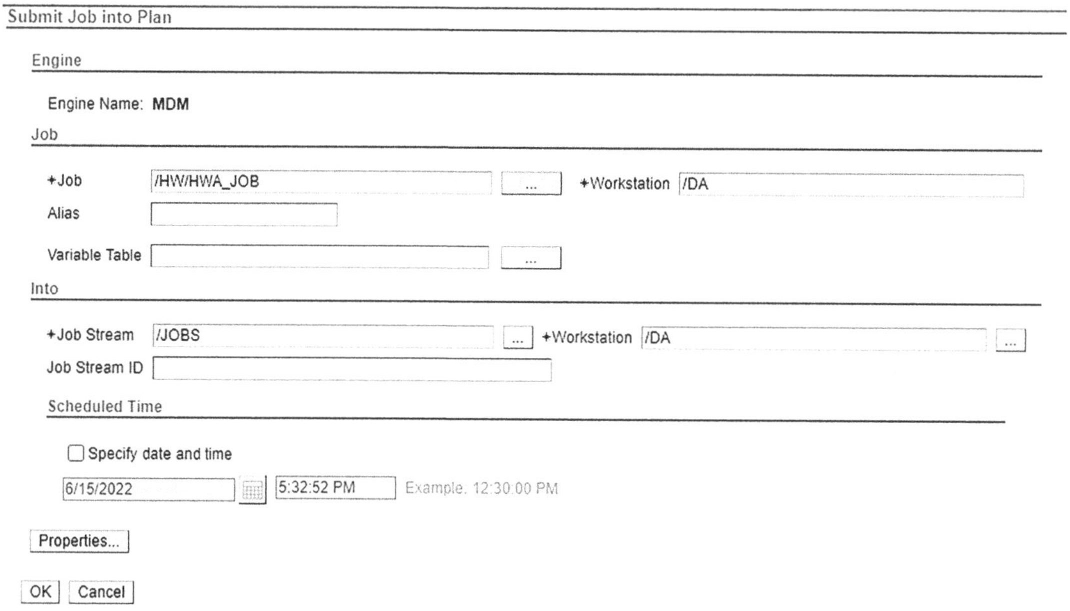

Submit Job into Plan

Engine

Engine Name: **MDM**

Job

+Job   /HW/HWA_JOB   [...]   +Workstation /DA

Alias

Variable Table   [...]

Into

+Job Stream   /JOBS   [...]   +Workstation /DA   [...]

Job Stream ID

Scheduled Time

☐ Specify date and time

6/15/2022   [▦] 5:32:52 PM   Example. 12:30:00 PM

[Properties...]

[OK] [Cancel]

Figure 4-34. *Submit Job Properties*

- Properties tab is used to submit the job on different conditions, in case the submitted job needs to be executed on specific timing or job dependency. Please refer to Figure 4-35 for properties options.

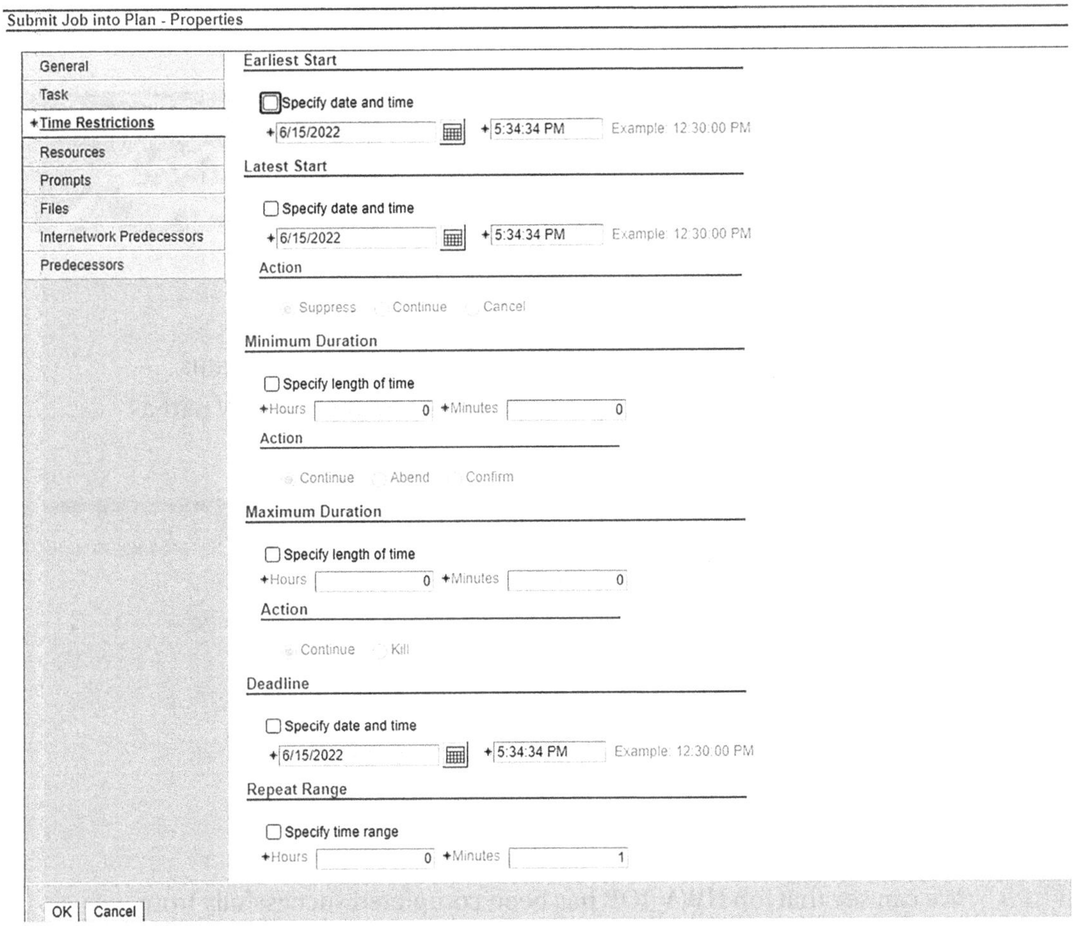

Submit Job into Plan - Properties

	Earliest Start
General	
Task	☐ Specify date and time
+ Time Restrictions	+ 6/15/2022 ▦ + 5:34:34 PM Example: 12:30:00 PM
Resources	Latest Start
Prompts	
Files	☐ Specify date and time
Internetwork Predecessors	+ 6/15/2022 ▦ + 5:34:34 PM Example: 12:30:00 PM
Predecessors	Action

Action

  ○ Suppress   ○ Continue   ○ Cancel

Minimum Duration

☐ Specify length of time
+ Hours [0] + Minutes [0]
Action

  ○ Continue   ○ Abend   ○ Confirm

Maximum Duration

☐ Specify length of time
+ Hours [0] + Minutes [0]
Action

  ○ Continue   ○ Kill

Deadline

☐ Specify date and time
+ 6/15/2022   ▦   + 5:34:34 PM   Example: 12:30:00 PM

Repeat Range

☐ Specify time range
+ Hours [0] + Minutes [1]

[OK] [Cancel]

Figure 4-35. *Advanced Option in Submit Job Properties*

- Once the job is submitted, we can monitor the job. To monitor the jobs in DWC, navigate to Monitoring and Reporting ➤ Monitor workload.

Figure 4-36. *Monitoring Tab for Job*

- The following interface opens where you can check the job status, select object type as job, and query the job name under Query part as shown in Figure 4-37 and select "Run."

Figure 4-37. *Object-Type Selection*

- We can see that job HWA_JOB has been completed successfully from Figure 4-38.

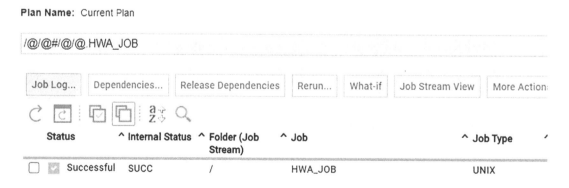

Figure 4-38. *Job Status View*

- We can check the job log for the jobs as well and select job log option as shown in Figure 4-38. Please see Figure 4-39 for job log.

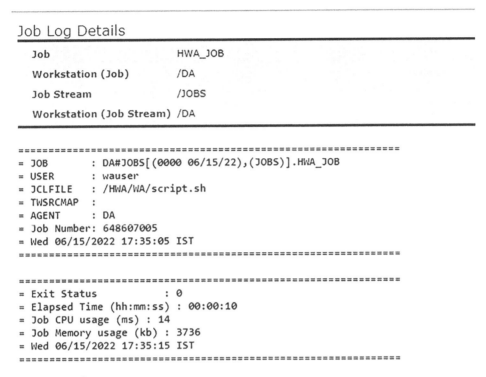

Job Log Details

Job	HWA_JOB
Workstation (Job)	/DA
Job Stream	/JOBS
Workstation (Job Stream)	/DA

```
==============================================================
= JOB      : DA#JOBS[(0000 06/15/22),(JOBS)].HWA_JOB
= USER     : wauser
= JCLFILE  : /HWA/WA/script.sh
= TWSRCMAP :
= AGENT    : DA
= Job Number: 648607005
= Wed 06/15/2022 17:35:05 IST
==============================================================


==============================================================
= Exit Status          : 0
= Elapsed Time (hh:mm:ss) : 00:00:10
= Job CPU usage (ms) : 14
= Job Memory usage (kb) : 3736
= Wed 06/15/2022 17:35:15 IST
==============================================================
```

Figure 4-39. *Job Log*

We have successfully submitted a predefined job on the Dynamic Workload Console and monitored the job status and job log.

We can now submit the predefined job stream. To submit the predefined job stream, navigate to Planning ➤ Submit Predefined Jobstreams.

Figure 4-40. *Submit Predefined Job Stream*

- The following interface opens where you can search for the job stream that needs to be submitted. In our case, the job stream name is "TEST_JS." Please refer to Figure 4-41.

Submit Job Stream into Plan

Select an Engine

Engine name MDM (Distributed) ⌄
Job Stream

+Job Stream /TEST_JS   ...   +Workstation /DA
Alias

Variable Table
Scheduled Time

☐ Specify date and time
6/15/22   5:39 PM   Example 12 30 PM

[Properties...]

[Submit]

Figure 4-41. *Submit Predefined Job Stream Properties*

- Properties option is same as predefined job submission, and select "Submit" to submit the job into the plan.

- Once the job stream is submitted, navigate to Monitoring & Reporting ➤ Monitor workload ➤ Object type ➤ Job stream as Figure 4-42, and query the job stream name in Query section as TEST_JS and select "Run."

Monitor Workload

Engine:	Object Type:	List Plans:
PRD_MDM	Job Stream	current-plan

Query:
/@/@/@/TEST_JS   [Run] [Edit] [View As Report]

Figure 4-42. *Monitoring View of Job Stream*

- The following interface opens where you can check the status of the job stream; in our case, the job stream is completed successfully. Please check Figure 4-43.

Figure 4-43. *Job Stream Status*

We have successfully submitted predefined jobs and job stream. And we can monitor their status as well from Dynamic Workload Console.

Use case 2:

We have a business requirement to submit a script on HWA agent where the job is not predefined in the database.

The requirements are as follows:

```
Script Path: /HWA/WA/script.sh
Server name: DA
Username: wauser
```

- Using Submit ad hoc job option, we can submit the jobs without creating this in HWA database. Using this option, we can submit the commands that need to be run on an agent machine like "ls -lrt" or "pwd."

To submit the job in ad hoc basis, navigate to planning ➤ submit ad hoc job. Please see Figure 4-44 for reference.

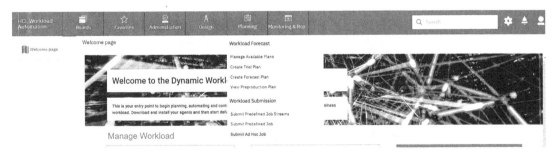

Figure 4-44. *Submit Ad Hoc Job*

- The following interface opens where you can update the username, server name, and job name for the script to be submitted. See Figure 4-45 for reference:

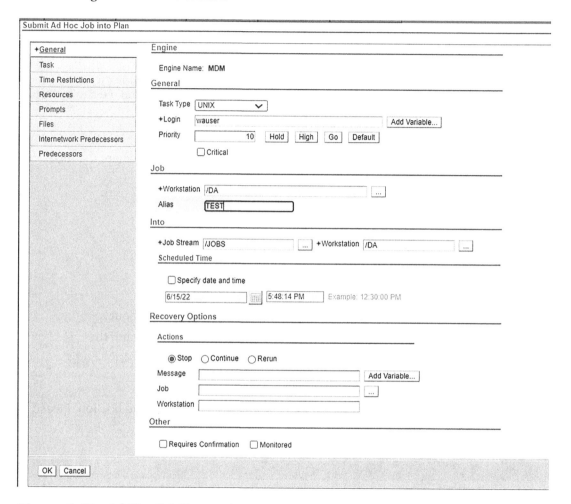

Figure 4-45. *Ad Hoc Job Properties*

- Please note that we don't have the job name from requestor, so we are giving the job name as "TEST."

- Update the script path on task properties. This is the executable that needs to run on an agent machine as per the requestor as shown in Figure 4-46. Alternatively, this can also be a command that needs to be run on an agent machine like "ls -lrt" or "pwd."

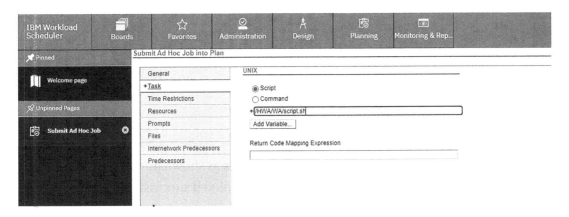

Figure 4-46. *Submit Ad Hoc Job Task*

- Once we update the script path, select ok to submit the job. Once the job is submitted, we can monitor the job under monitor workload object ➤Monitor job ➤ Object type as job as Figure 4-47.

Figure 4-47. *Status of Ad Hoc Job*

- We can see that job got successfully submitted and completed.

Use case 3:

We have a request to rerun the job HWA_JOB1 on server DA. Please follow the following steps to rerun the job.

To rerun the job in current plan, navigate to Monitoring and Reporting ➤ Monitor workload ➤ Object type as job and query the job name as HWA_JOB1 as shown in Figure 4-48 and select "Run."

Figure 4-48. *Monitoring Job*

- This opens the interface to view the job status as shown in Figure 4-49, and we can see that job got completed successfully. To rerun the job, select Rerun option as shown in Figure 4-49.

Figure 4-49. *Job Status View*

- This Rerun option opens the interface and reruns the job as shown in Figure 4-50. Select Rerun to rerun the job.

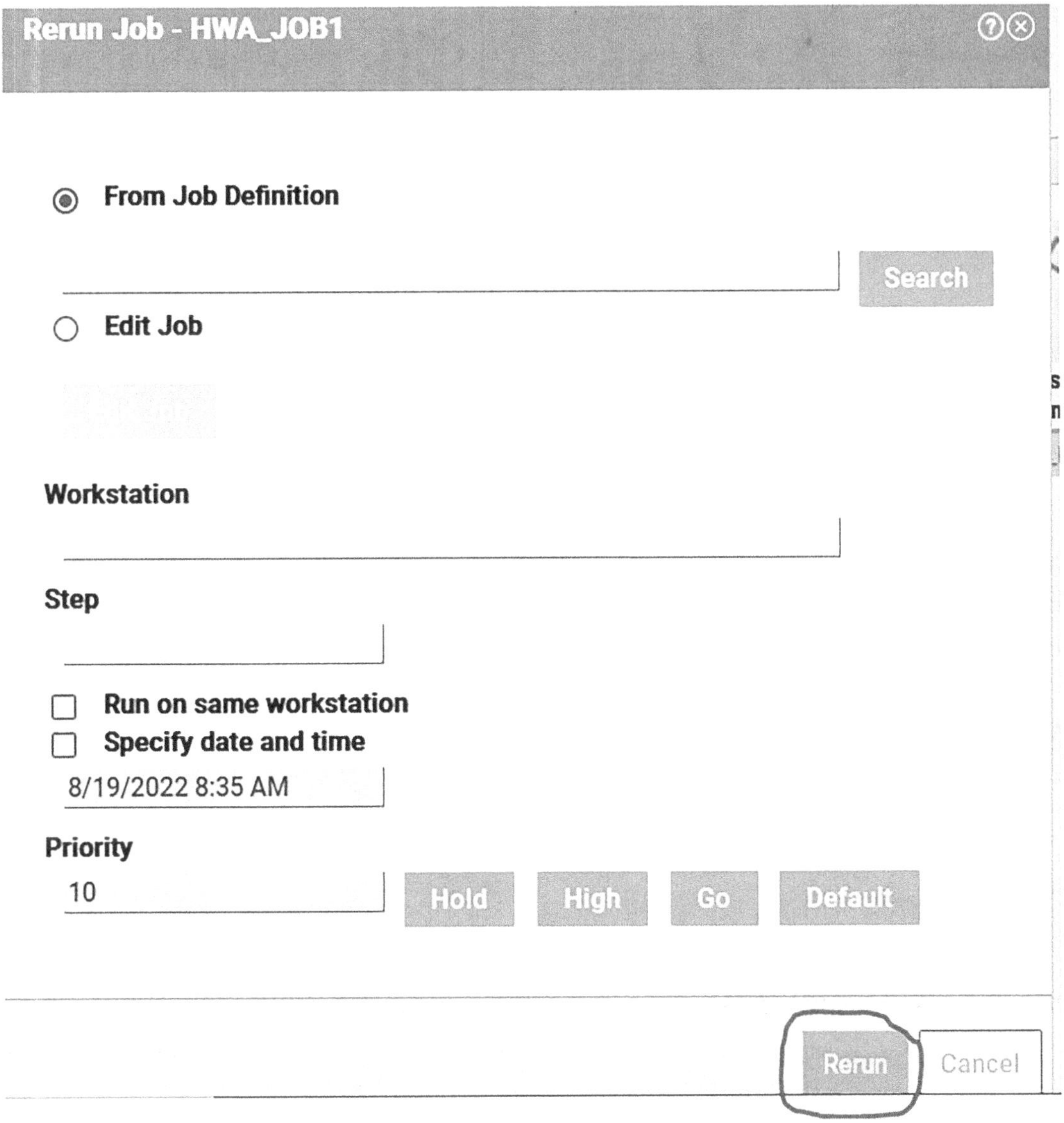

Figure 4-50. *Rerun Option*

- We could see that job got successful after rerun; please look at the actual start time after rerun to confirm that rerun has performed as shown in Figure 4-51.

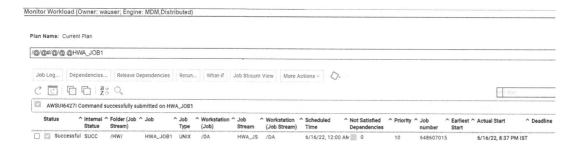

Figure 4-51. *Rerun Option*

We have successfully rerun the job as per the request.

Use case 4:

We have a requirement to cancel the job HWA_JOB2. Cancelling a job prevents the job to be started and lets successors to start; if we want to prevent successors to start immediately, we can use "cancel pending" that actually releases successors only when the cancelled job would be ready to start.

To cancel the job from DWC, navigate to Monitoring and Reporting ➤ Monitor Workload ➤ Object type as job. Query the job name as HWA_JOB2 in query section as shown in Figure 4-52.

Figure 4-52. *Monitoring Job*

- This opens the interface to view the job status as shown in Figure 4-53, and we can see that job got completed successfully.

Figure 4-53. *Job Status View*

- To cancel the job, select the job and select more actions ➤ cancel option as shown in Figure 4-54 and select "Cancel."

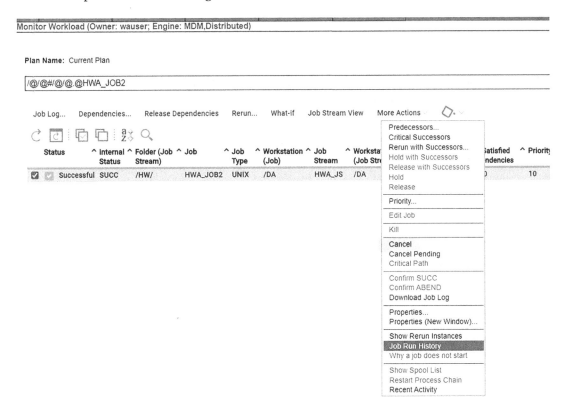

Figure 4-54. *More Actions View*

- We can see that the job got cancelled successfully as shown in Figure 4-55.

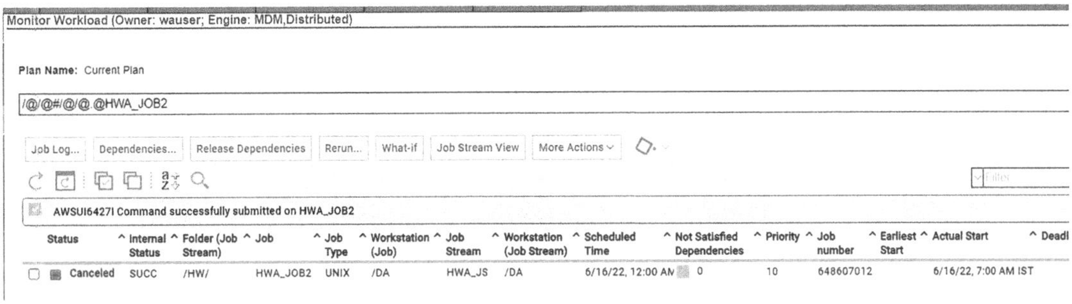

Figure 4-55. *Cancelled Job Status*

We have successfully cancelled the job as per the request.

95

Summary

In this chapter, we covered the following capabilities of HWA:

- How to create a job and job streams

- How to submit an ad hoc job in HWA

- How to rerun and cancel the jobs in HWA

CHAPTER 5

HWA for Managed File Transfers

In this chapter, we will focus on the managed file transfer feature of HWA which provides a technology platform to reliably exchange data between various systems in a secured manner and adhere to the compliance needs.

HCL workload automation provides the following security protocols to enable secured file transfers from source to destination:

- FTP (File Transfer Protocol)

- FTPS (File Transfer Protocol Secure)

- FTPES (Explicit File Transfer Protocol)

- SSH (Secure Shell)

To transfer the files to destination server, HCL workload automation agent (dynamic agent or fault tolerant agent) has to be installed on the Source server. This is prerequisite for any type of file transfers mentioned above.

Business Requirement

Let us consider a business scenario, where data needs to be exchanged between various business processes during the execution of a business process workflow like in a case where organizations credit the salary to their employees once in a month; we can relate to the following business processes:

- Payroll batch job generates the salary amount of every employee for a particular month and feeds in the data sheet and stores it in the salary database.

© Navin Sabharwal and Subramani Kasiviswanathan 2023
N. Sabharwal and S. Kasiviswanathan, *Workload Automation Using HWA*,
https://doi.org/10.1007/978-1-4842-8885-6_5

- Data sheet is a data file that needs to be securely transferred to the destination bank data server. This file is transferred to destination server by secured FTP jobs in HWA.

- The transferred file is processed for payment at a scheduled time using HWA's scheduling feature.

- Post payment SMS is triggered to the respective registered mobile numbers using the integration feature of HWA.

- The next business process or HWA job runs and generates the pay slips.

- The last job in HWA triggers secured file transfer jobs to send pay slips to all the employees.

  To technically understand how to configure a file transfer job, follow the following steps. In the above real-time workflow, let us learn how to configure the business process that sends the salary data sheet (two sheets) to the bank's data server. Source server: svaliapwlap001.dryicelabs.com

  ```
  Files to be transferred: /home/wauser/Salary_DataSheet.xlsx
                            /home/wauser/Salary_DataSheet1.xlsx
  Protocol: SSH
  Job name: HWA_SSH_DATASHEETS_MONTHLY
  HWA server name: /MDM_DA
  Destination server: svaliapwlap002.dryicelabs.com
  Destination path: /home/wauser
  ```

The requirement is to create a file transfer job to send the files using MDM_DA agent server, and the protocol has to be SSH. Please follow the following steps to create the preceding job.

- Log in to Dynamic Workload Console and navigate to Design ➤ Manager Workload Definition and select Engine as shown in Figure 5-1.

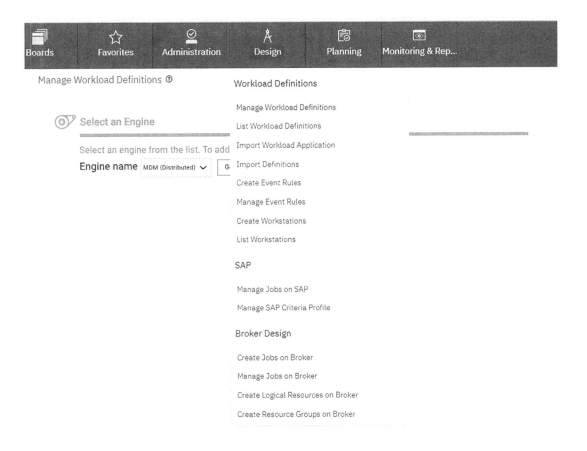

Figure 5-1. *DWC Home Page View*

- This opens the following interface where you can create the file transfer job. Select Job Definition as shown in Figure 5-2.

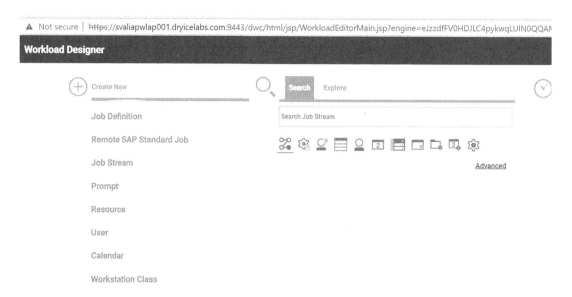

Figure 5-2. *Job Definition Page*

- This opens the following page and you have to search for file transfer
  plug-in job and select the File Transfer job as shown in Figure 5-3.

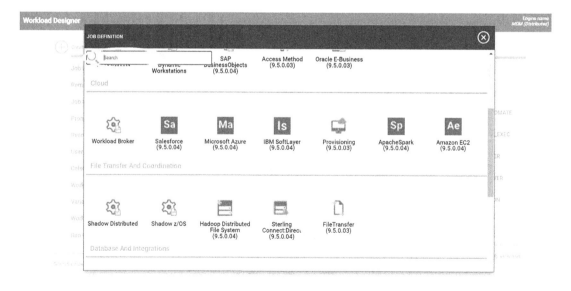

Figure 5-3. *FTP Plug-In Image*

- Update the job name and server name as per the requirement. In our case, Job Name: HWA_SSH_DATASHEETS_MONTHLY and Server Name: /MDM_DA is shown in Figure 5-4.

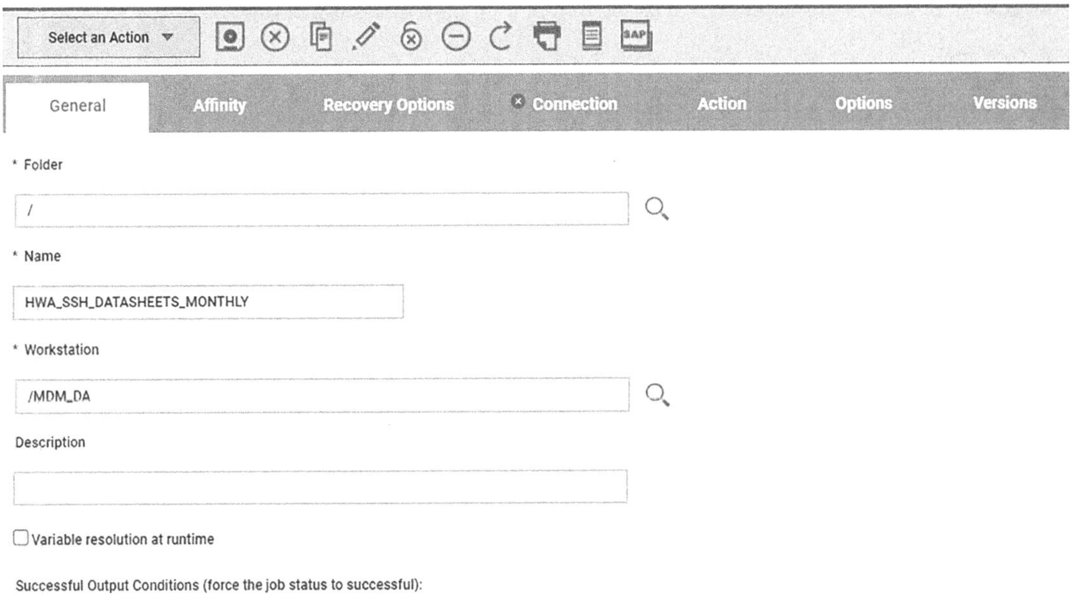

Figure 5-4. *Job Definition Properties*

- Update the Connection details under Connection tab. As per the request, select the Connection settings as "SSH" shown in Figure 5-5.

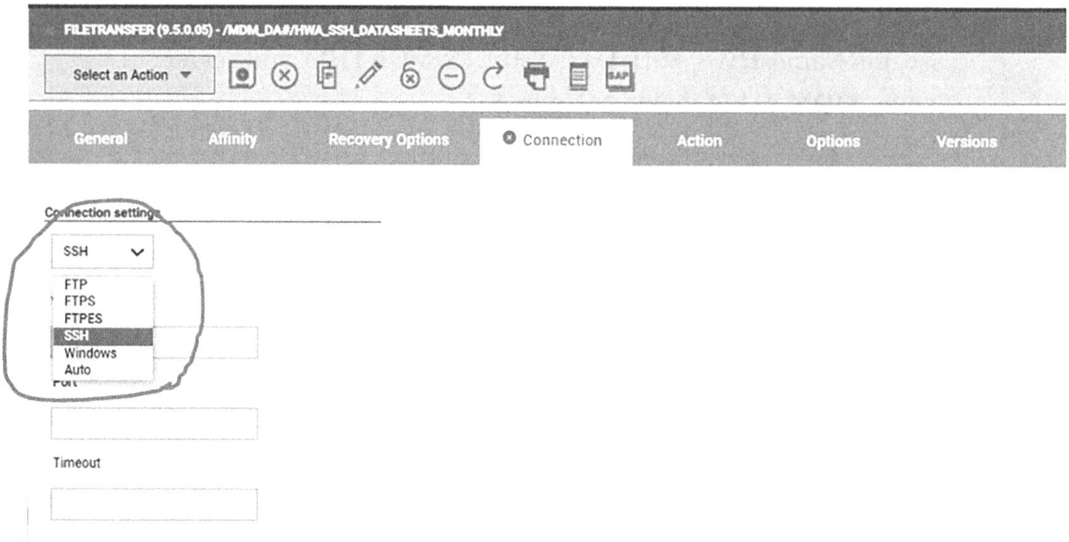

Figure 5-5. *Job Definition Connection Properties*

- In the Credentials option, we need to update the Destination server User and Password for authentication purpose. See Figure 5-6:

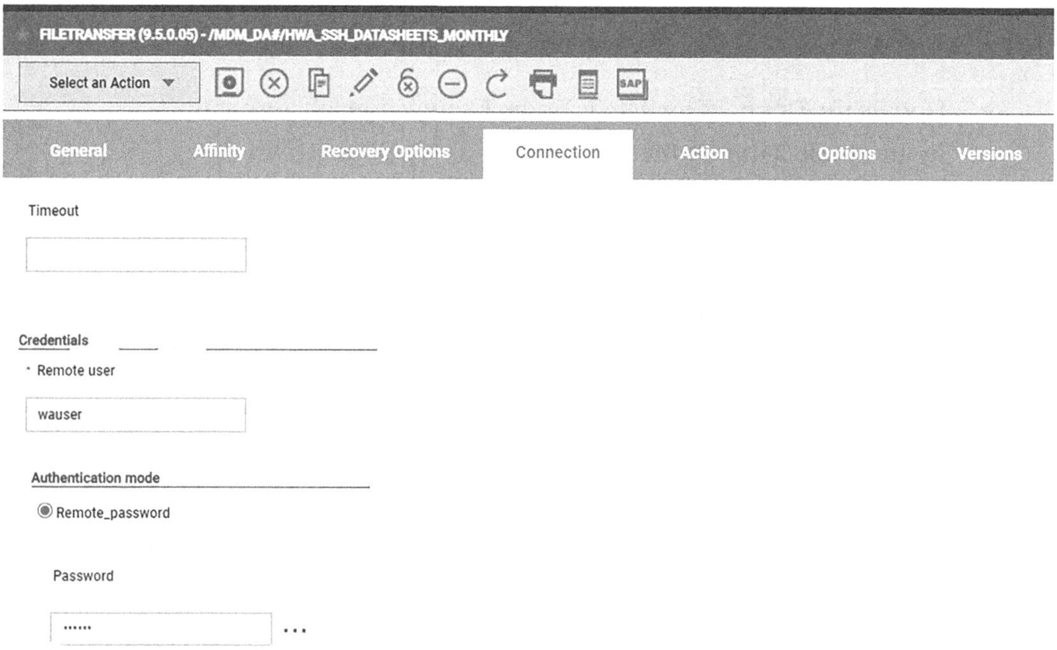

Figure 5-6. *Job Definition Connection Properties*

- Keystore option is available in HWA; in case if the requirement is to have a keystore password, we can update that password in keystore section which provides enhanced security.

- Select "Test Connection," and make sure the connection is successful to send the file, as shown in Figure 5-7:

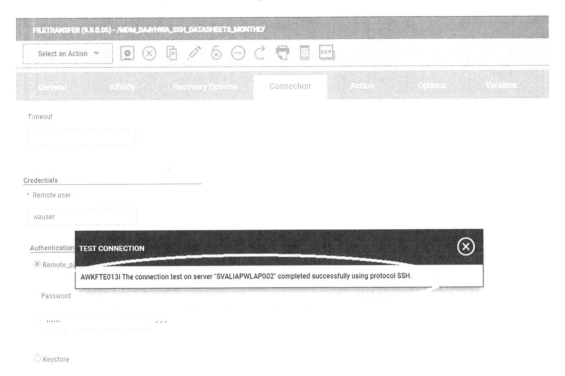

Figure 5-7. *Test Connection Results*

- Once the connection is successfully tested, we can update the source and destination files. Refer to Figure 5-8 and Figure 5-9 for reference.

- In our case, the source files are /home/wauser/ Salary_DataSheet. xlsx and /home/wauser/ Salary_DataSheet1.xlsx. Refer to Figure 5-8:

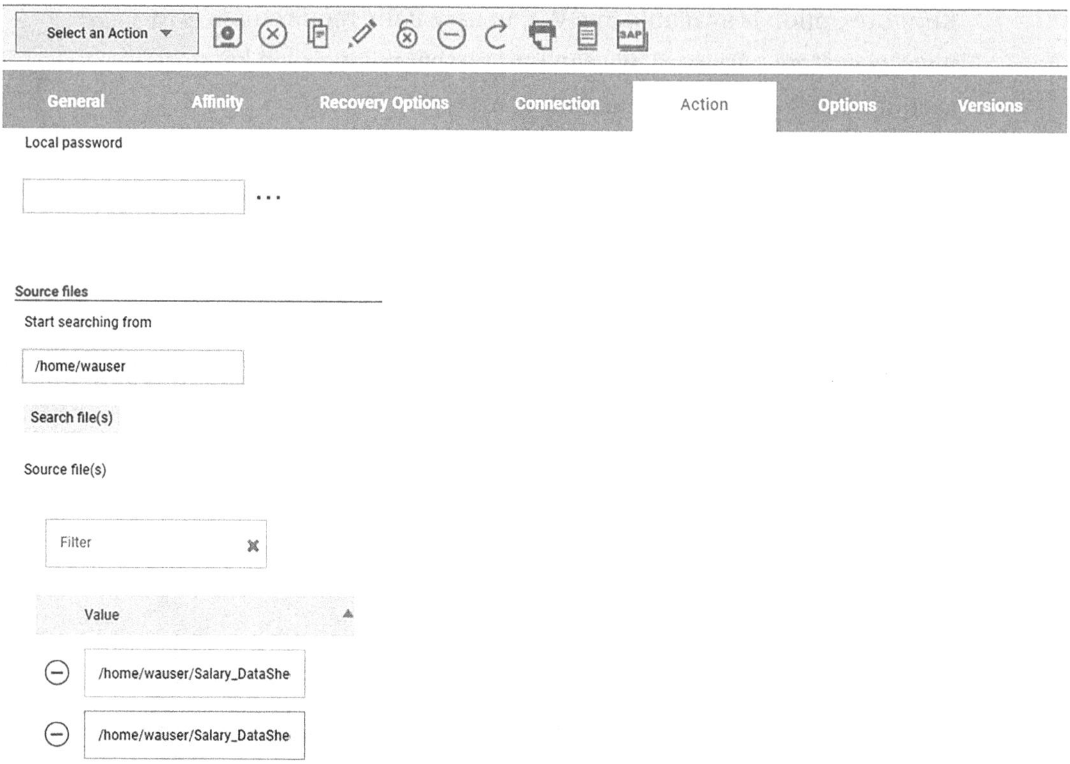

Figure 5-8. *Source File Information*

- The Destination path is /home/wauser, as you can see in Figure 5-9.

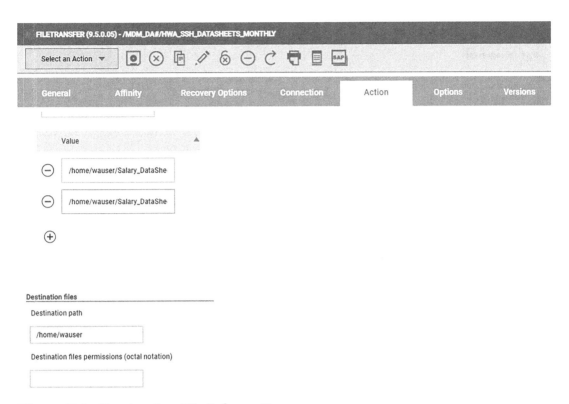

Figure 5-9. *Destination File Information*

- In options tab, HWA has additional transfer options. We can Replace, Rename, Append, Skip, or Abend the files if in case it is already present on the destination server. In our case there is no such requirement; so as shown in Figure 5-10, we are choosing default option "Replace."

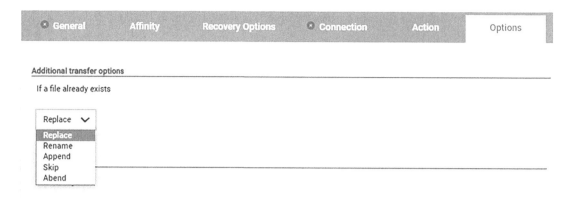

Figure 5-10. *Destination File Information*

- Select the Transfer mode as per the requirements. We can select Binary/Text method. In our case, it is only two files to be sent to destination server, so we are going with default option "Binary" as shown in Figure 5-11.

Transfer mode

◉ Binary

○ Text

Source file codepage

Destination file codepage

Wait for file timeout

Maximum simultaneous transfers

FTP options

Connection mode

◉ Passive

○ Active

Min port

Max port

Figure 5-11. *Transfer and FTP Options*

- In Connection mode, default setting would be Passive to select system setting and we can select Active type as well. In Active mode, the server enables the connection with the client using min and max ports. The Active connection type is selected when the user wants to send/receive the data in dedicated ports. In our case, it is Passive, as you can see in Figure 5-12:

As shown in Figure 5-12, we have successfully created the job.

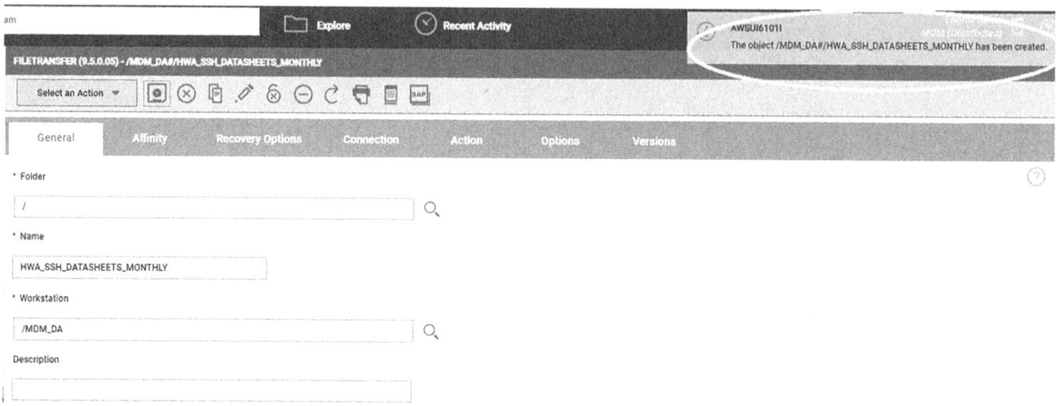

Figure 5-12. *Job Creation Completed*

Since the job has been created, we can now submit the job and validate if the file transfer has been done automatically using HWA. To submit the job, navigate to Select an Action ➤ Submit the job into Current plan and select "Submit" as shown in Figure 5-13:

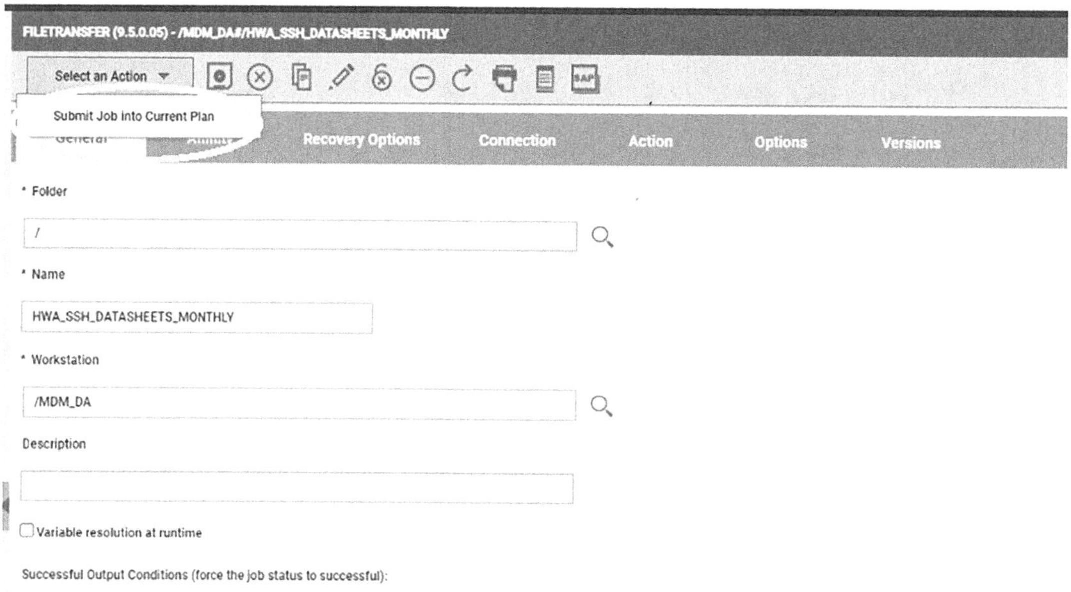

Figure 5-13. *Job Submission*

Check the job log and validate if the file has been transferred to destination server from source server. See Figure 5-14:

```
[wauser@SVALIAPWLAP002 ~]$ pwd
/home/wauser
[wauser@SVALIAPWLAP002 ~]$ ls -l MFTbeta_readme.pdf
-rw-rw-r--. 1 wauser wauser 108827 Jan 20  2022 MFTbeta_readme.pdf
```

Figure 5-14. *Job Log Information*

- We could see from the preceding log that the job got completed
  successfully and files have been transferred to destination server. We
  can log in to destination server and check whether the files have been
  transferred.

- Log in to destination server; in our case we have to log in to
  svaliapwlap002.dryicelabs.com server and change the directory to /
  home/wauser and check for the files as shown in Figure 5-15.

wauser@SVALIAPWLAP002:~

```
[wauser@SVALIAPWLAP002 ~]$ pwd
/home/wauser
[wauser@SVALIAPWLAP002 ~]$ ls -l
total 128
-rw-rw-r--. 1 wauser wauser      76 Jun 22 23:06 1
drwxrwxrwx. 6 root   root       59 Jun 20 20:27 images
drwxr-xr-x. 2 wauser wauser      93 Feb 22 17:11 javasharedresources
-rw-rw-r--. 1 wauser wauser 108827 Jan 20 18:46 MFTbeta_readme.pdf
drwxrwxrwx. 2 root   root     4096 Jun 20 20:25 plugins
-rw-rw-r--. 1 wauser wauser      76 Jun 22 23:06 Salary_DataSheet1.xlsx
-rw-rw-r--. 1 wauser wauser      78 Jun 22 23:0  Salary_DataSheet.xlsx
-rwxr-xr-x. 1 wauser wauser    2273 Feb 28 18:06 storage_api.py
[wauser@SVALIAPWLAP002 ~]$
```

Figure 5-15. *Destination Server Files*

- We can see that files are transferred successfully to destination server.

Summary

In this chapter, we covered the following capabilities of HWA:

- Capabilities of HWA in terms of managed file transfers.

- To create, submit and validate a file transfer job via a real-time
  business use case.

CHAPTER 6

HWA Integration with SAP Application

In this chapter, we will be discussing about the working principle of SAP job integration with HCL workload automation. SAP stands for Systems Applications and Products in data processing.

SAP products are used in manufacturing, life science, distribution, and technical and business services. The top industries that use SAP business are healthcare, energy, education, telecommunications, finance, retail, etc.

SAP R3batch access method enables communication between external SAP systems with HCL workload automation and provides a single point of entry for automating and launching the jobs. Using this method, we can run SAP process chain and ABAP jobs. As a prerequisite, HWA agent (dynamic agent or FTA) has to be installed on SAP application server to run the SAP jobs.

An enterprise scheduler like HWA is put to use at its best when there is a need to interact and exchange data between multiple systems which are a combination of SAP and non-SAP that are interdependent and are running their services in different platforms like on-prem, hybrid, cloud, etc. If the use case is to just use SAP systems and jobs are only restricted to SAP, then there is no need for an external workload scheduling tool.

HWA has capabilities to automate execution of business processes for all the SAP modules like the following:

SAP PIChannel

SAP BusinessObjects

SAP Data Services

SAP Cloud Platform Integration for Data Services

SAP HANA Cloud Platform Application Lifecycle

© Navin Sabharwal and Subramani Kasiviswanathan 2023
N. Sabharwal and S. Kasiviswanathan, *Workload Automation Using HWA*,
https://doi.org/10.1007/978-1-4842-8885-6_6

SAP HANA Database

SAP Integrated Business Planning

SAP Hana Extended Application Services (XS)

SAP NetWeaver Java

Business Requirements

Let us consider a business scenario, where data need to be exchanged between various business processes during the execution of a workflow like, in a case where consumer goods companies process their invoices to different vendors. Let us summarize the process.

- Vendor generates invoices and stores it in invoice database of the consumer goods company.

- Financial department validates the invoices for different vendors and processes the invoices using SAP application. In SAP, there are many modules created to execute the business workflow based on their requirement.

- To process the invoices based on the departments, we can use HWA job and generate the bills.

- The generated bills can be sent to all the vendors directly using HWA SAP jobs based on the applications.

    To technically understand how to configure SAP jobs, follow the following steps. In the above workflow, let us learn how to configure the SAP jobs in HWA and send bills to the vendors.

    ```
    Server name            : ABC
    SAP job name           : SAP_ABAP
    SAP job type           : SAP ABAP
    ```

    Requirement is to create a SAP ABAP job to send bills to the vendors across the consumer goods company. Please follow the following steps to create the above job.

Prerequisites:

- The below library files have to be installed on the agent under <TWSDATA/methods> directory.

Operating System	Unicode SAP NetWeaver RFC SDK libraries
AIX 64-bit	libsapucum.so
	libsapnwrfc.so
	libicuuc50.a
	libicui18n50.a
	libicudecnumber.so
	libicudata50.a
Linux.64	libicuuc.so.50
	libsapucum.so
	libicudata.so.50
	libicui18n.so.50
	libsapnwrfc.so
	libicudecnumber.so

Figure 6-1. *Library Files*

- From SAP end, RFC user has to be created with CPIC, Communications, or DIALOG attributes.

- Create the authorization profile with PFCG profile generator in SAP system.

- On your SAP database server, log on to the SAP system as an administrator and copy the control file and data file from the methods directory to the following directories on your SAP database server:

- copy control_file /usr/sap/trans/cofiles/

- copy data_file /usr/sap/trans/data/

- The names of control_file and data_file vary from release to release. The files are in the methods directory – UNIX®: TWA_DATA_DIR\ methods – and have the following file names and format:

- For SAP releases 6.10 or later:

  K900044.TV1 (for standard jobs scheduling)
  R900044.TV1 (for standard jobs scheduling)
  K900751.TV1 (for IDoc monitoring and job throttling)
  R900751.TV1 (for IDoc monitoring and job throttling)

- SAP Transport Request contains two files: First one starts with "R" character and second one stars with "K" character.

First one that starts with "R" character should be placed in server folder: "/usr/sap/trans/data/"

And second one that starts with "K" character should be placed in server folder: "/usr/sap/trans/cofiles/"

- Once we have completed the above prerequisites, we have to create the r3batch files for HWA to SAP connections. To create the r3batch file, we would need R3 (SID, client, instance, host, user, and password) details to connect SAP application. We would need to collect these details from SAP application and create an <agent>_r3batch.opts file under </TWSDATA/methods> directory as shown in Figure 6-2.

```
[wauser@SVALIAPWLAP002 methods]$ cat ABC_XA_r3batch.opts
r3host=ABC.domain.com
r3sid=ABC
r3client=200
r3auditlevel=2
r3instance=00
r3user=HCL_HWA
RETRY=1
```

Figure 6-2. *R3batch Options File*

- Crete the following workstation definition to execute the jobs from ABC_XA server:

```
CPUNAME ABC_XA
  OS OTHER
  NODE null TCPADDR 31111
  TIMEZONE GMT
  FOR MAESTRO HOST ABC_FTA ACCESS "r3batch"
    TYPE X-AGENT
    AUTOLINK OFF
    BEHINDFIREWALL OFF
    FULLSTATUS OFF
END
```

- Once the workstation is created, we have to test the connection to SAP system using "./r3batch -t PL -c ABC_XA -l * -j * -- "-debug -trace" command, and the output of the command should be as you see in Figure 6-3.

```
%job "/SAP/BATCHJOB_TESTING  " HCL_HWA     10553700 01/21/2022 16:34:53 C

%job "/SAPAPO/BATCH          " HCL_HWA     05021500 01/21/2022 16:36:24 C

%job "/SAPAPO/DELETE         " HCL_HWA     05024900 01/21/2022 16:25:34 C

%job "/SAPAPO/TEST_DELETE        " HCL_HWA     06293400 01/21/2022 16:25:34 C

%job "/SAPAPO/BATCH_PDRP1        " HCL_HWA     08564100 01/21/2022 16:37:13 C

%job "/SAPAPO/MODID          " HCL_HWA     19235800 01/21/2022 16:34:22 C

%job "/SAPAPO/TS _CHECK      " HCL_HWA     05021500 01/21/2022 16:37:57 C

%job "/SAPAPO/TS_LCM_CHECK      " HCL_HWA     19253300 01/21/2022 16:37:57 C

%job "/SAPAPO/TS_CHECK_R    " HCL_HWA     00283500 01/21/2022 16:36:24 C
```

Figure 6-3. *R3batch Connection Output*

- Once we get the output as shown in Figure 6-3 we can confirm that HWA to SAP connection is successful and we can create the job. Please find the following procedure to create and submit the job.

- To create the job definition, navigate to Design ➤ Manage Workload Definition and select Engine. Refer to Figure 6-4 for reference.

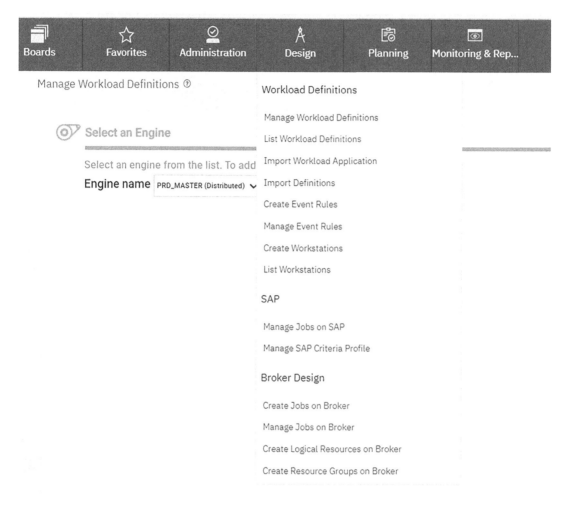

Figure 6-4. *Design Page*

- This opens the interface where we have to select job definition type as "SAP job on XA Workstation" to create the job and select the plug-in to create the job. Refer to Figure 6-5 for reference.

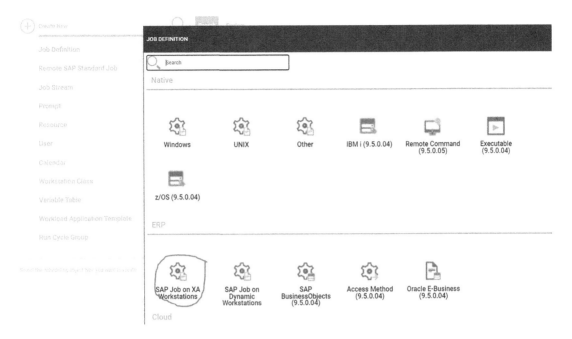

Figure 6-5. *SAP Job Plug-in*

- This opens the interface to update the job name, server name, and username to run the job; in our case, the job name is SAP_ABAP and server name is ABC_XA, as shown in Figure 6-6.

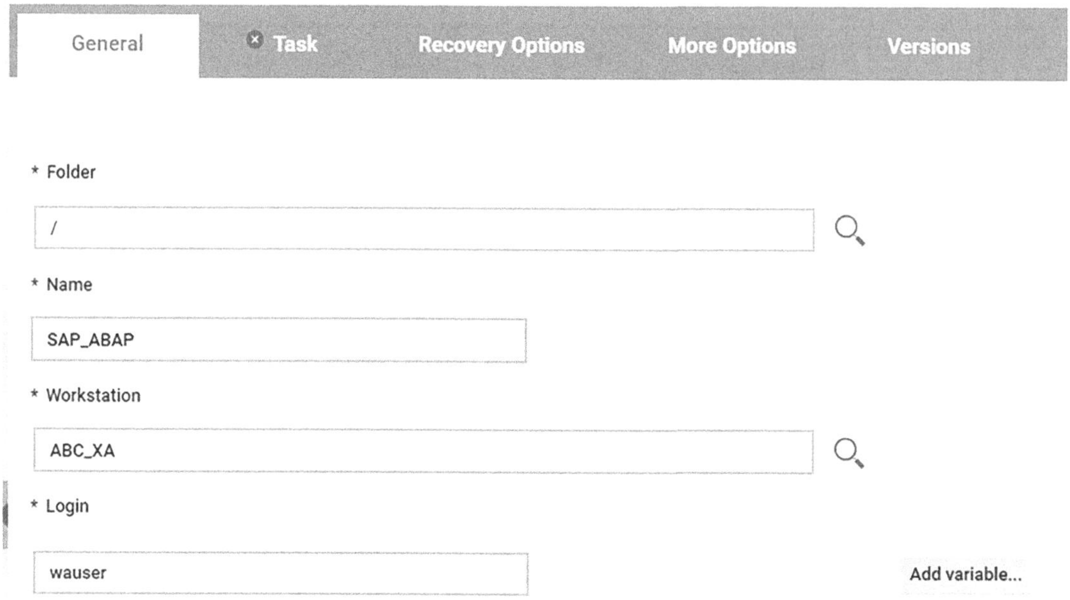

Figure 6-6. *SAP Job General Properties*

- Under task section, the subtype has to be standard since we are
  creating SAP ABAP job. In case of process chain job, we have to select
  process chain type. Search for SAP job under search section and
  select the job name as shown in Figure 6-7 for reference and save
  the job.

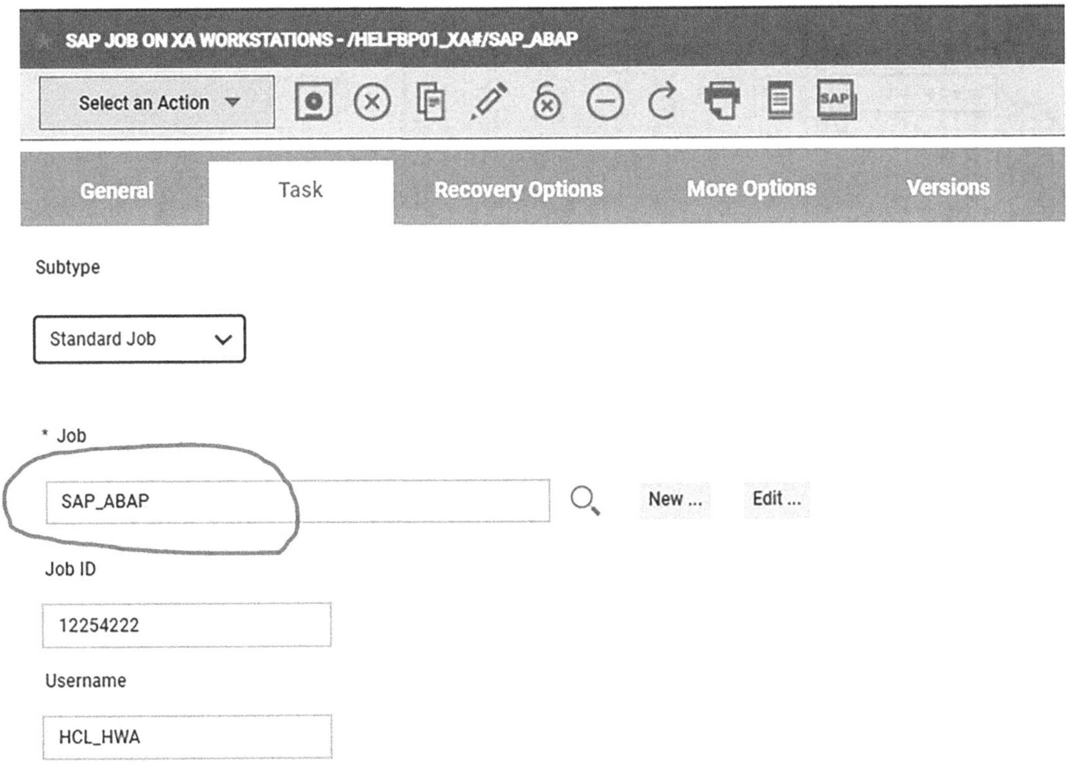

Figure 6-7. *SAP Job Properties*

- Once the job is created, select "select an action" tab and submit the job. Refer to Figure 6-8 for reference.

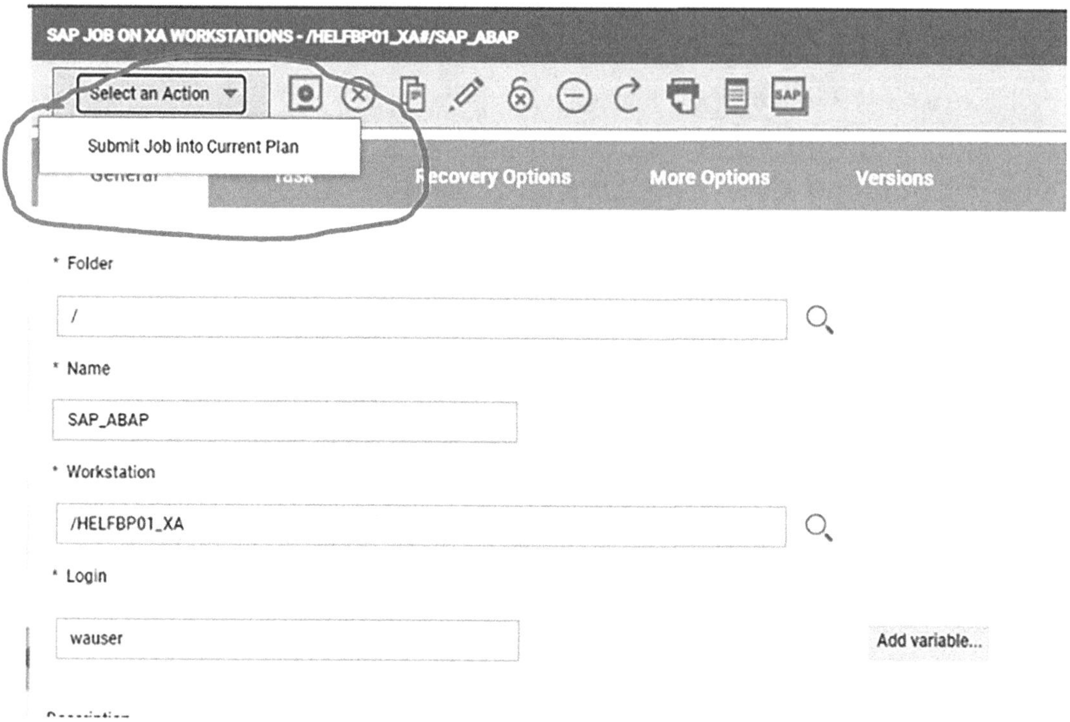

Figure 6-8. *SAP Job Submission Option*

- Once the job is submitted, navigate to Monitoring and Reporting ➤ Monitor workload. Please check Figure 6-9 for reference.

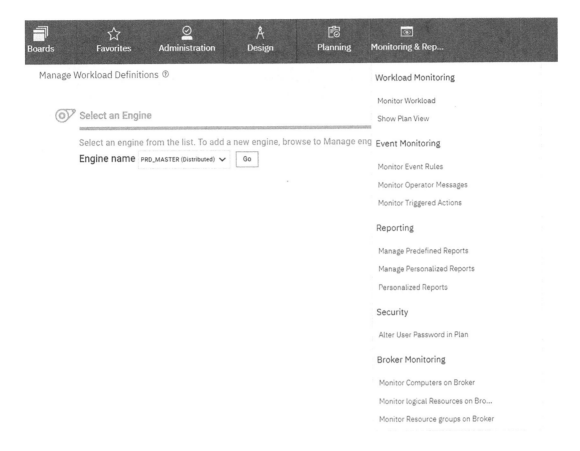

Figure 6-9. *Monitor Workload Option*

- Select the object type as "job" and query the job name SAP_ABAP to check the status of the job and select "Run." Refer to Figure 6-10.

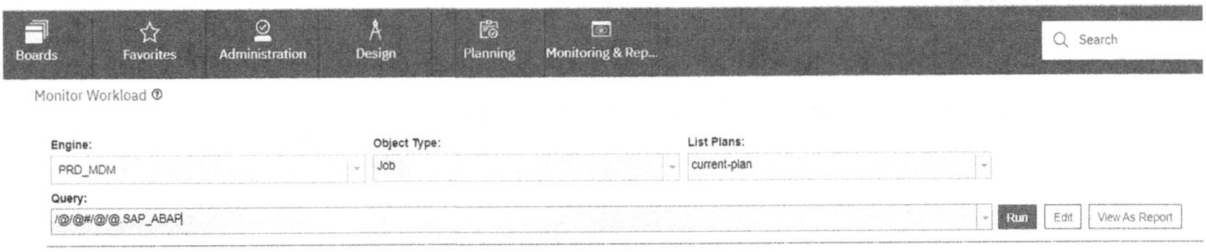

Figure 6-10. *Object-Type Selection*

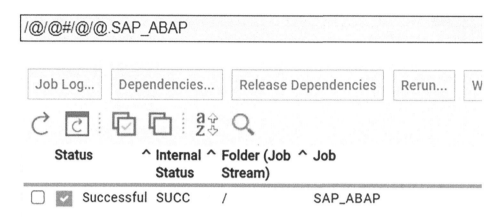

Figure 6-11. *SAP Job Status*

We could see that the job got completed successfully as shown in Figure 6-11, and we can create SAP process chain jobs as well using same method we need to create the job and schedule it based on the requirement.

As per our requirement, we have successfully created the job SAP_ABAP and it got completed successfully.

Summary

To summarize the content of this chapter, we covered the following:

- Capabilities of HCL workload automation to automate the business processes for SAP systems.

- Technically understand and configure SAP plug-in and create, design, and execute jobs to run on SAP systems.

CHAPTER 7

Automate Job Executions on Microsoft SQL Server

In this chapter, we are going to discuss about HCL workload automation integration with MS-SQL server jobs. MS-SQL job performs specific set of action on MS-SQL server agent. Single MS-SQL job can run on local server or multiple remote servers.

Microsoft SQL server is a relational database management system (RDBMS) and supports a wide variety of transaction processing, business intelligence, and analytics applications. With HCL workload automation, we can automate the MS-SQL jobs to run on the MS-SQL target machines. These jobs are used to update, insert, and retrieve data from a database. There are various other use cases where SQL server jobs are used.

Business Requirement

Let us consider a real-time use case where the customer wants to run their MS-SQL query to generate a KPI for the specific month. Here is the brief process:

- MS-SQL query generates the KPI report for the specific month on KPI database. The output of the query will be sent to respective functional team for their analysis.

- Every time customer has to log in to MS-SQL server to generate the reports, this process consumes time and manual work is involved.

- We can integrate MS-SQL server with HWA and execute the MS-SQL query as an automated job.

  Let us proceed to create a simple MS-SQL job with HWA. In the preceding workflow, let us create the MS-SQL plug-in job.

© Navin Sabharwal and Subramani Kasiviswanathan 2023
N. Sabharwal and S. Kasiviswanathan, *Workload Automation Using HWA*,
https://doi.org/10.1007/978-1-4842-8885-6_7

```
Job name            : MSSQL_KPI
HWA server name     : MDM_DA
MS-SQL Db server    : SVALIAPWLAP002@dryicelabs.com(10.1.152.227)
Port number         : 1433
MS-SQL job name     : DB-KPI-Master
Username&Password   : db_user&XXXX
```

Please follow the following steps to create the job in HWA:

- Navigate to Design ➤ Manage Workload definition ➤ Select Engine, as shown in Figure 7-1.

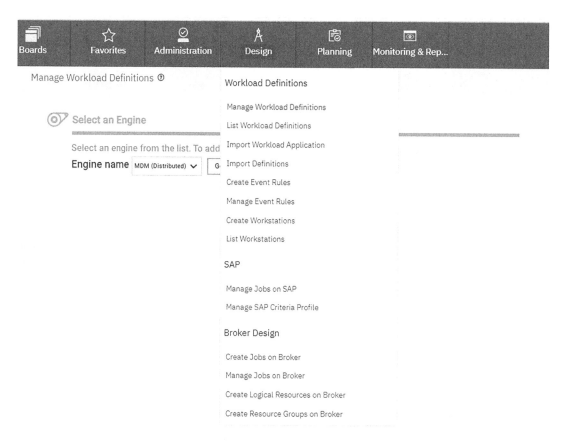

Figure 7-1. *DWC Home Page View*

- This opens the following interface where you can create the MS-SQL job. Select Job definition as shown in Figure 7-2.

Navigate to Create New ➤ Search for MS-SQL job and select MS-SQL job plug-in. Again, see Figure 7-2.

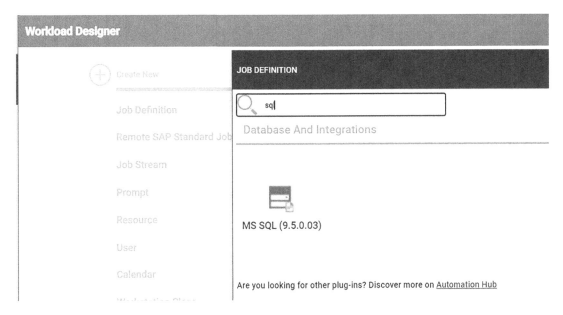

Figure 7-2. *MS-SQL Job Plug-in View*

- This opens the following interface to update the MS-SQL job details; in the General tab, select the required dynamic agent server and mention the job name.

- In our case, the job name is MSSQL_KPI and HWA server name is MDM_DA. Refer to Figure 7-3:

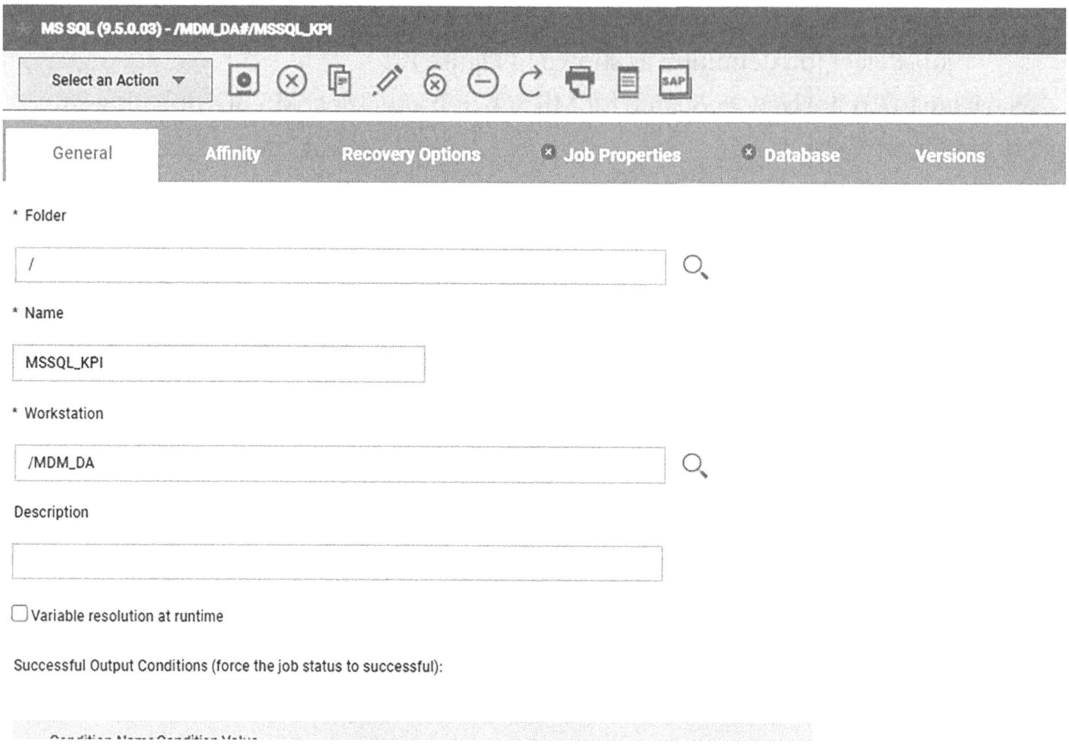

Figure 7-3. *HWA Job Properties*

- Before updating the job properties, Under Database tab, update the following information shown in Figure 7-4 to connect to MS-SQL server and get the job name.

- Database name.

- JDBC jar class path.

- Database server and port.

- Database credentials.

- Make sure the test connection is successful. If the test connection is not successful, we need to check the database credentials and that the database is available and reachable.

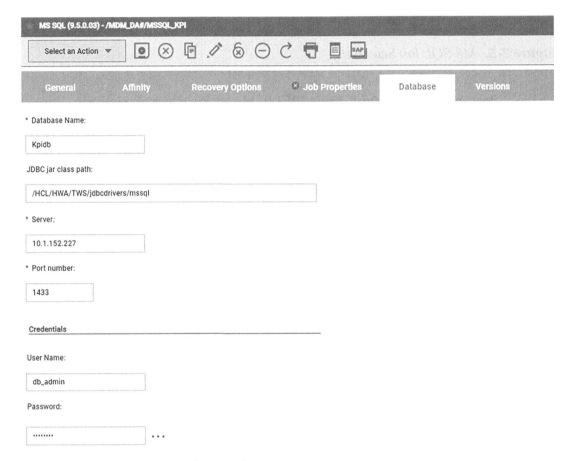

Figure 7-4. MS-SQL Database Selection

- Once the connection is successful, select "Get Jobs" under Job properties tab to get the job name from MS-SQL server. Refer to Figure 7-5.

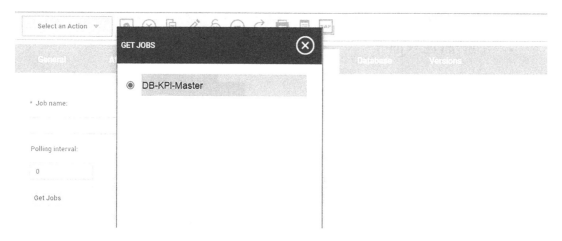

Figure 7-5. *MS-SQL Job Selection*

- We can see from Figure 7-5 that DB-KPI-Master job is already cleared in MS-SQL server. Select DB-KPI-Master job and select "ok."

- Readers need to note that the actual job is created in SQL Server; we are discovering it here and running it from HWA. For creating SQL server jobs, please refer to SQL server documentation.

- Save the job definition and submit the job and validate the status of the job. To submit the job, navigate to Select an Action ➤ Submit the job into Current plan and select "Submit." Please check Figure 7-6 for reference.

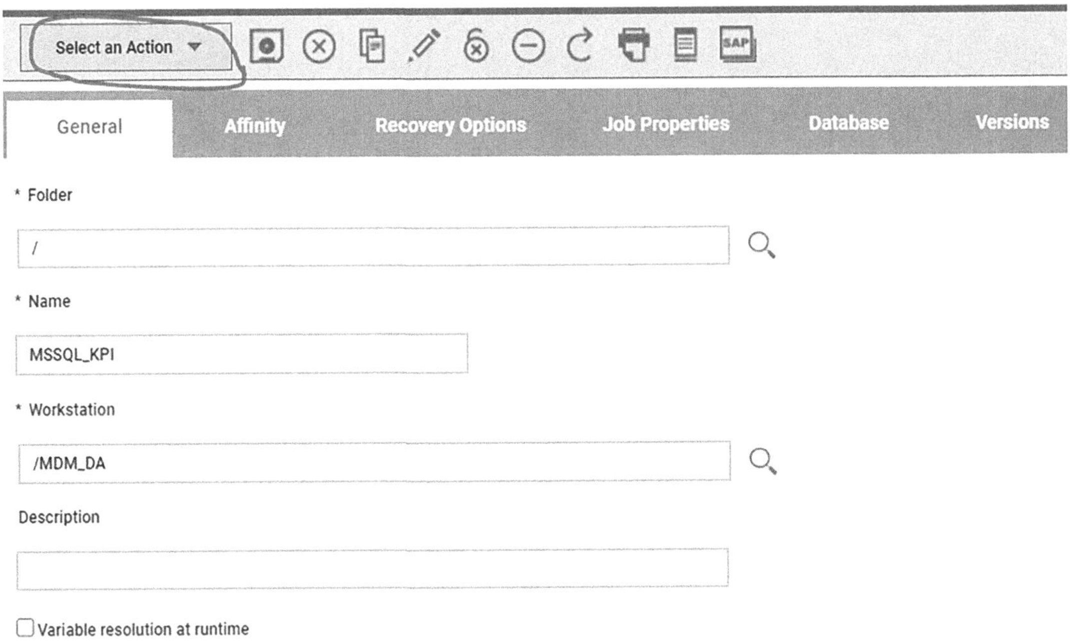

Figure 7-6. *Submit Job Option*

- We have successfully created the job and it completed successfully. Please see the following log for reference.

```
Job log:
        = Exit Status          : 0
        = Elapsed Time (hh:mm:ss) : 00:00:42
        = Wed 03/02/2022 21:02:17 MYT
```

- We have successfully imported an MS-SQL job and automated using HCL workload definition. Readers can use other advanced features of HWA to schedule and create complex scenarios with multiple SQL server jobs as explained in earlier chapters.

Summary

To summarize the content of this chapter, we covered the following:

- Integrating HWA with SQL server

To discover, submit and validate a MS-SQL job via HWA.

CHAPTER 8

Working with RESTful Web Services

This chapter will cover REST APIs in HWA. REST API is a mediator between the users, clients, resources, and web services. REST APIs interact with the HWA system to retrieve information and perform actions and communicate with the system. API is used to share resources and their information while maintaining security, control, and authentication to determine access control details of the user.

HWA provides the following methods to interact with the service URL and process the actions based on the user selection. Detailed explanation of REST APIs and their usage is beyond the scope of this book; readers are expected to refer to online sources on REST API usage.

- GET
- POST
- PUT
- HEAD
- DELETE

In HWA, we can schedule jobs that add, download, delete, and modify resources or data on RESTful web services via HTTP methods in any available content type such as JSON, XML, and XHTML.

© Navin Sabharwal and Subramani Kasiviswanathan 2023
N. Sabharwal and S. Kasiviswanathan, *Workload Automation Using HWA*,
https://doi.org/10.1007/978-1-4842-8885-6_8

Business Requirement

Let us consider a use case where we want to get the status of a web URL in a text file.

- Customer has a dedicated service URL to perform transactions.

- The service URL is published on the Internet, and users access the URL all the time.

- We want to check the availability of the service URL `https://petstore3.swagger.io` all the time.

- To monitor the service URL, we can create a REST API job using method "GET"; please follow the following steps to get the output of the service URL.

```
Job name            : TEST_GET_PETSTORE_DATA
Server name         : MDM_DA
Service URI         : https://petstore3.swagger.io
Method              : GET
Output file path    : /home/wauser/TWS/test.txt
Username            : wauser
```

- Log in to Dynamic Workload Console and navigate to Design ➤ Manage workload definition as depicted in Figure 8-1:

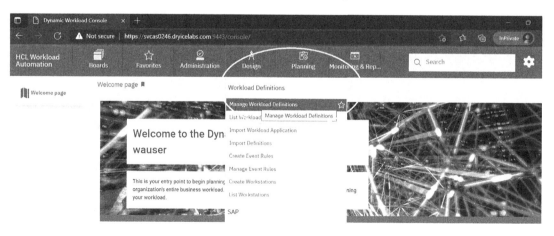

Figure 8-1. *DWC Home Page*

- This opens the interface to create the new job definition. Select the Restful job definition as shown in Figure 8-2 and update the required fields.

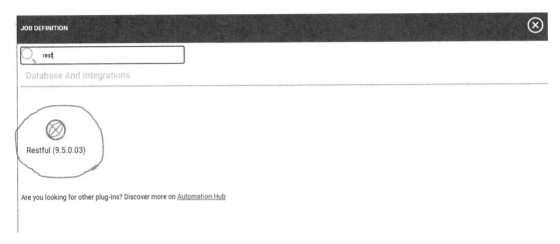

JOB DEFINITION

rest

Database And Integrations

Restful (9.5.0.03)

Are you looking for other plug-ins? Discover more on Automation Hub

Figure 8-2. *Restful Job Plug-in Page*

- Update the Rest api job name and name of the dynamic agent where the job will execute. Please note that we can execute restful job using dynamic agent (DA) only and we can't execute the rest api job using fault tolerant agent (FTA). In our case, the job name is TEST_GET_PETSTORE_DATA and server name is MDM_DA. Refer to Figure 8-3.

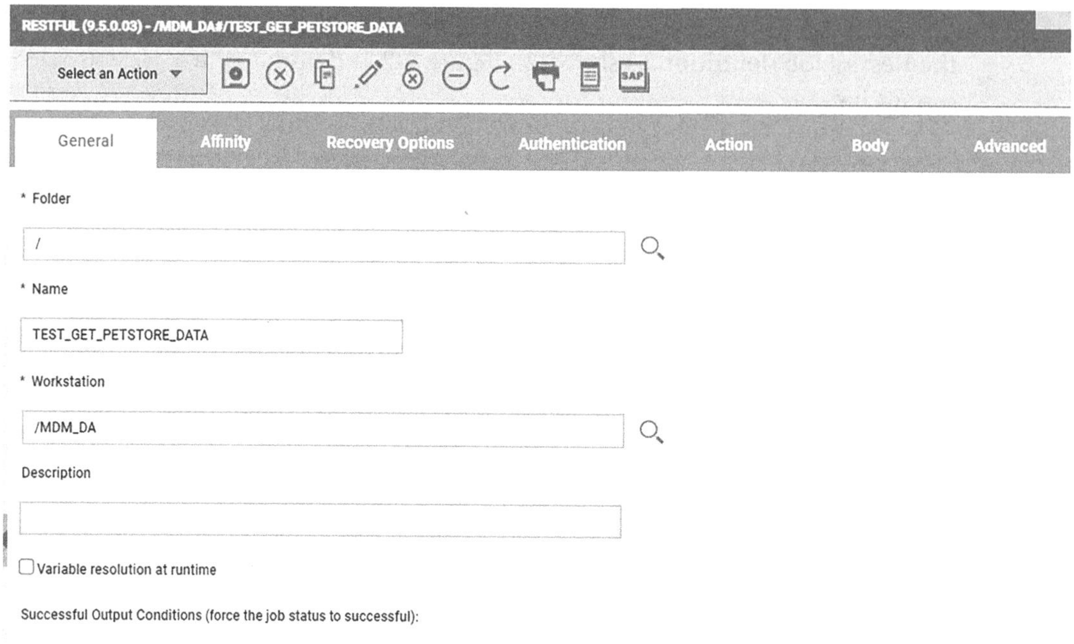

Figure 8-3. *Job Properties Section*

- Select the authentication and update the required credentials for the URL. In case we are using certificates instead of clear text password, we can update the keystore path as mentioned in Figure 8-4. In our case the credentials are wauser.

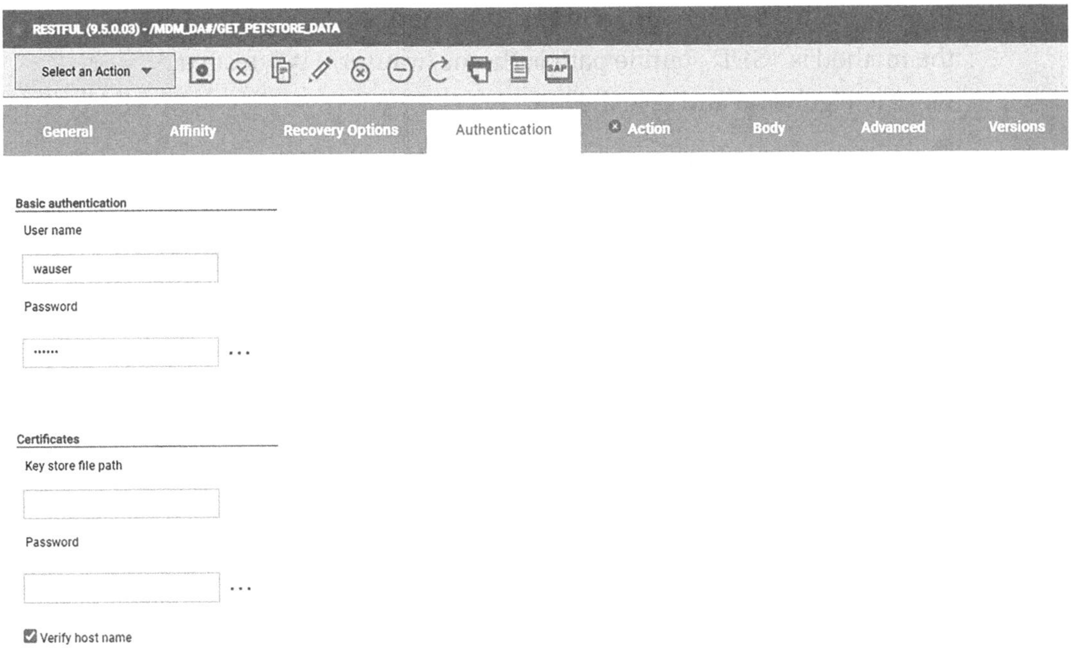

Figure 8-4. *Authentication Details*

- Under Action tab, update the service URL and select the method that is required to run this job. If the output of the job must be stored in any file, then mention the same in output filename. The available methods are as follows:

- GET – retrieves the header and body of a resource.

- POST – creates a resource.

- PUT – changes the state of a resource or updates it.

- HEAD – retrieves only the header of a resource without its body.

- DELETE – removes a resource.

- Output file name – enter the full name of the file where the response from the web resource is to be returned.

- Http query parameters – enter one or more query parameters that tailor and filter the response output. Query parameters are case-sensitive.

- In our case, the service URL is `https://petstore3.swagger.io`, and the method is "GET"; outfile path is /home/wauser/TWS/test.txt. Refer to Figure 8-5 and save the job.

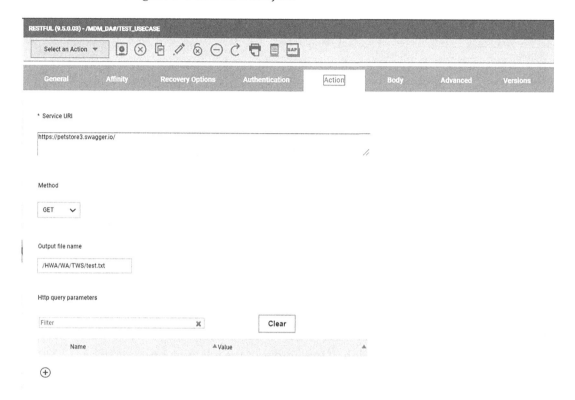

Figure 8-5. *REST API Action Tab*

The following is the explanation for each feature available in HWA rest api job.

- Body section is used to enter the body of the request.

- In body tab, we have content-type option where we can specify the media type of the entity-body sent to the recipient resource. In the case of HEAD method, update required media type.

- Body input option is used to update the body directly. In body input method, we can directly update the body or we can update the path of body to get it updated while the job execution is happening.

- Under Advanced tab, we can update additional parameters for rest api jobs. Http headers are available under advance tab option to update one or more HTTP or custom headers associated with your HTTP request. HTTP headers define the operating parameters of the transaction (metadata).

- Accept option is used to specify the media type in which the response from the web resource is to be returned. This type is usually specified by the resource to which you are sending the request.

- Number of retries – the maximum number of retries, in case of connection failure. Default value is 0.

- Body and advance options are used based on the requirement.

- We have successfully created the job and we can submit the job to test the execution of the job. To submit the job, navigate to Select an Action ➤ Submit Job into Current Plan. Refer to Figure 8-6.

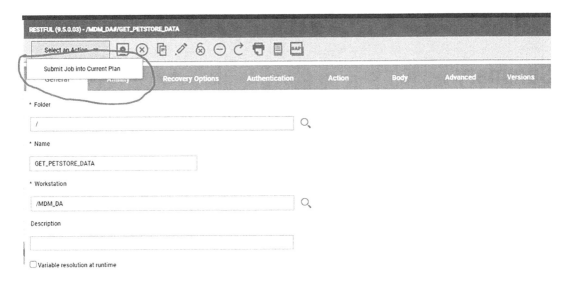

Figure 8-6. *Submit Action Tab*

Once the job is submitted, please check the job status in the Dynamic Workload Console. To check the status of the job, navigate to Monitoring and Reporting ➤ Monitor Workload ➤ Object type as jobs. You can see this in Figure 8-7:

Figure 8-7. *Job Status in Monitor Workload*

We can see that job got completed successfully. Please see the following job log.

Job log:
= Exit Status   : 0
= Elapsed Time (hh:mm:ss) : 00:00:02
= Thu 02/03/2022 18:22:19 JST

We can see that output of the service URL is copied to "/HWA/WA/TWS/test.txt" file. We have successfully created and tested a simple REST API job using HWA.

Summary

To summarize the content of this chapter, we covered the following:

- Capabilities of HWA in terms of Rest API jobs.

- To create, submit and validate a Rest API job.

## CHAPTER 9

Submit, Orchestrate, and Monitor Jobs on a Kubernetes Cluster

In this chapter, we will focus on Kubernetes job feature of HWA which provides easier way to perform the actions on Kubernetes from HCL workload automation. Kubernetes is used to manage containerized workloads and their services. By using Kubernetes job, we can submit and monitor the Kubernetes jobs on Kubernetes cluster. The Kubernetes batch job integration enables to run temporary containers via Kubernetes batch jobs and orchestrate and synchronize these containers between them.

To perform the actions on Kubernetes cluster, HWA dynamic agent has to be installed, and HWA application user should have access to the configuration file (e.g., $HOME/.kube/config). This is prerequisite for integration with Kubernetes. A deep dive on Kubernetes is beyond the scope of this book, and readers can use online resources to learn about Kubernetes and container orchestration.

Business Requirement

Let us consider a requirement where we want to deploy a new Pod on the Kubernetes environment. To technically understand how to configure Kubernetes batch jobs in HWA, follow the following steps. In the following workflow, let us learn how to deploy "mysql-deployment" using HWA. To create the Kubernetes job, we required the following details.

© Navin Sabharwal and Subramani Kasiviswanathan 2023
N. Sabharwal and S. Kasiviswanathan, *Workload Automation Using HWA*,
https://doi.org/10.1007/978-1-4842-8885-6_9

```
Job name             : KUBE_MYSQL_DEPLOYMENT   n
Server name          : MDM_DA
Config file path     : /HWA/WA/kube_config
Yaml/json file path: /k8s-o11y-workshop/mysql/mysql.yml
Namespace            : default
```

To create the job definition, navigate to Design ➤ Manage
Workload Definition and select Engine. Refer to Figure 9-1.

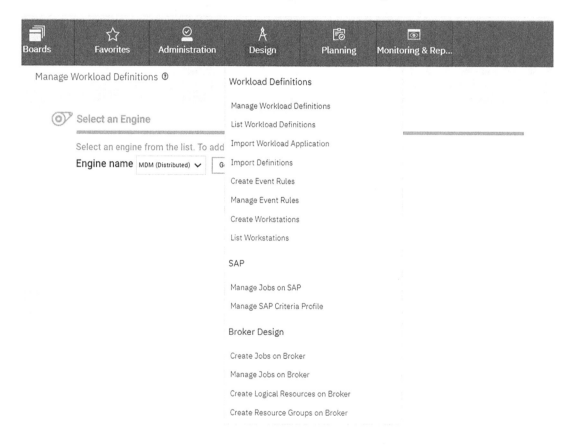

Figure 9-1. *DWC Home Page View*

This opens the following interface where you can create the Kubernetes batch job. Select Job definition and search for Kubernetes batch jobs, as shown in Figure 9-2.

JOB DEFINITION

kube|

Cloud

Kubernetes batch
jobs (10.1.0.00)

Figure 9-2. *Kubernetes Batch Job Plug-in*

Update the HWA job name and dynamic agent name where Kubernetes is installed. In our case, job name is KUBE_MYSQL_ DEPLOYMENT and dynamic agent name is MDM_DA. Refer to Figure 9-3.

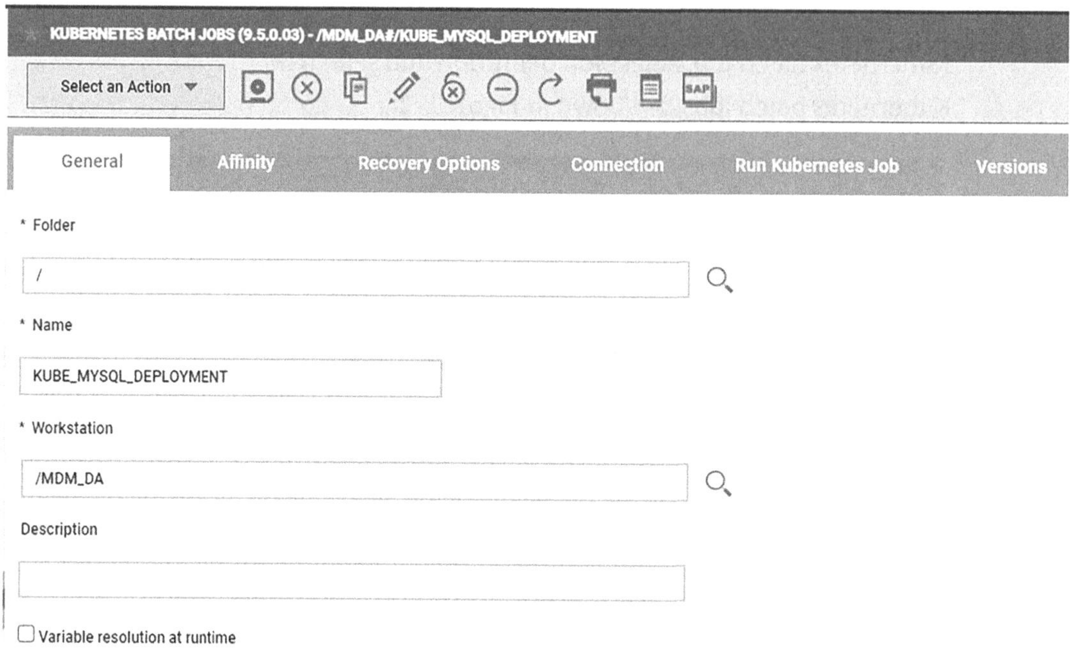

Figure 9-3. *Job Definition Properties*

Select connection tab and update Kubernetes server credential details. Refer to Figure 9-4:

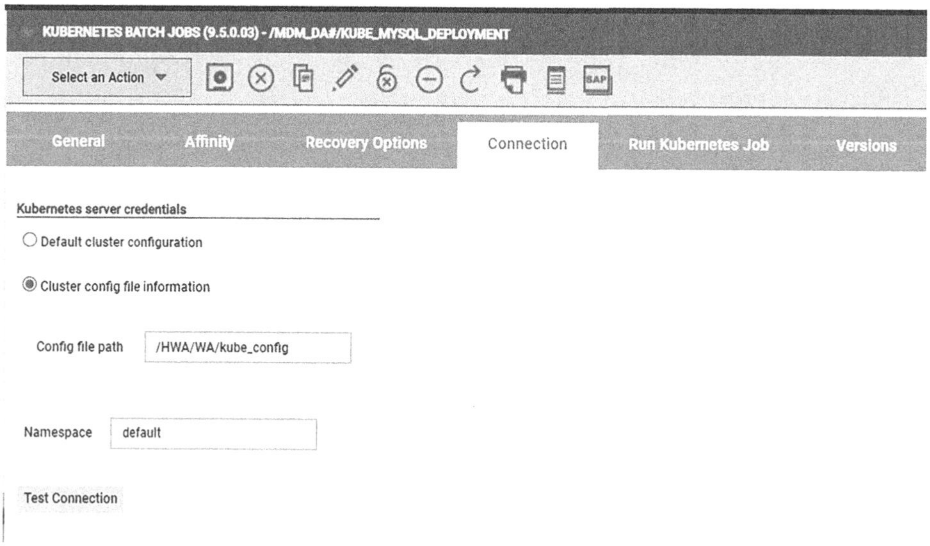

Figure 9-4. *Kube Connection Parameters*

Default cluster configuration – choose this option to establish a connection to the Kubernetes cluster if the agent is configured/deployed as a service in the K8's cluster. User can opt for "Default cluster configuration" in which the agent is preconfigured with cluster information.

Cluster config file information – choose this option to establish a connection to a remote K8's cluster. In this case, user needs to provide the path for the config file in the agent with the cluster information either in JSON or YAML format.

In case cluster config file information is selected, update the config file path and Namespace information as shown in Figure 9-4. In our case, config path is "/HWA/WA/ kube_config" and namespace is "default"; make sure test connection is successful.

Select Run Kubernetes job tab and update "submit and track a job" information. Only yaml/json formats are available in Kubernetes batch job of HWA.

In our case, we have given the yaml file path. Refer to Figure 9-5 and save the job.

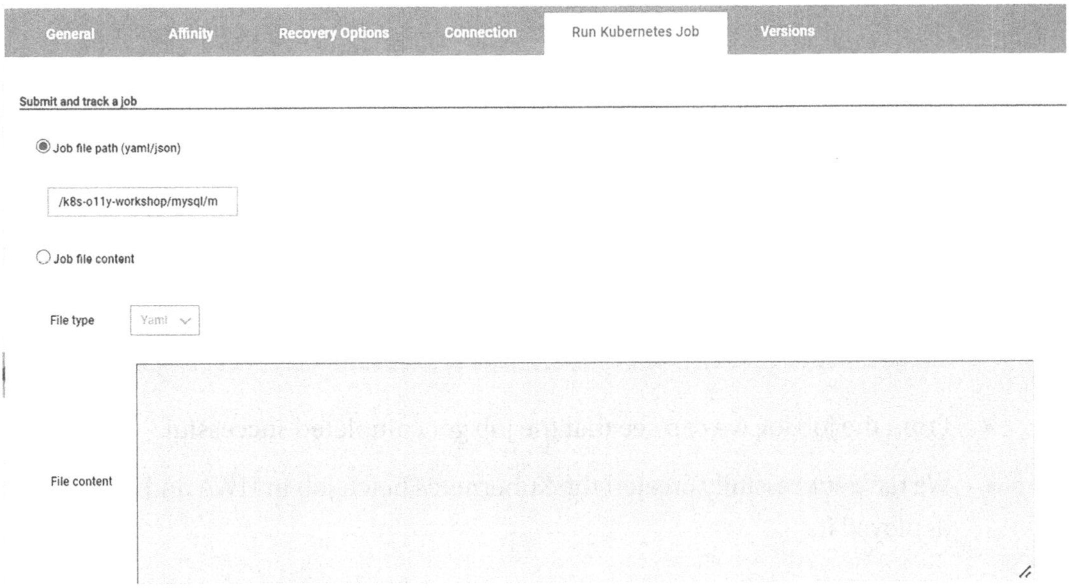

Figure 9-5. *Kube Process Information*

Once the job is created, submit the job in HWA. To submit the job, select "select an action" tab and submit the job. Refer to Figure 9-6 for reference:

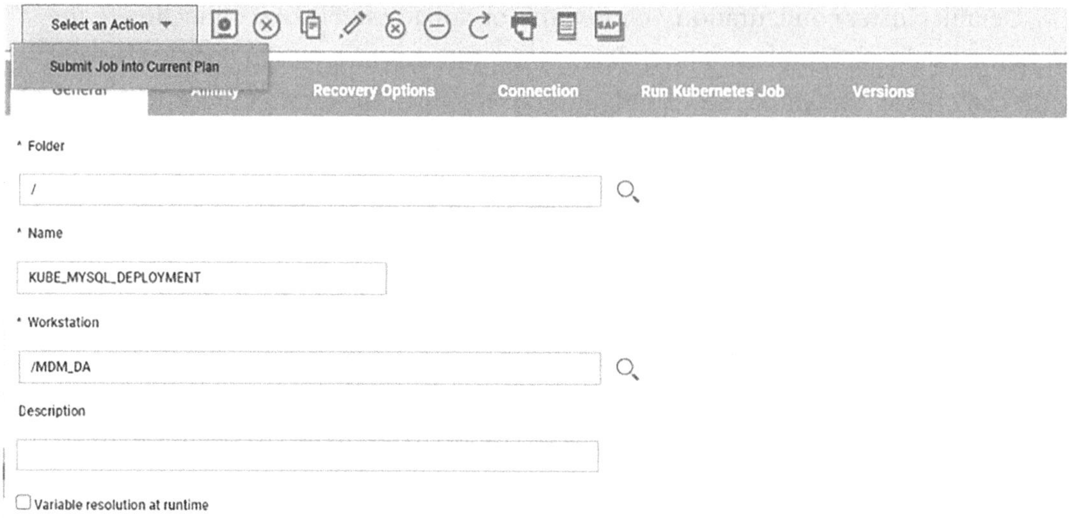

Figure 9-6. *Submit Job Option*

- The job got submitted successfully and please see the following
  job log.

```
==============================================================
= Exit Status           : 0
= Elapsed time (hh:mm:ss) : 00:00:12
= Wed 04/06/2022 19:42:50 JST
==============================================================
```

- From the job log we can see that the job got completed successful.

- We have successfully created the Kubernetes batch job in HWA and
  deployed it.

Summary

In this chapter, we covered the following capabilities of HWA:

- Capabilities of HWA in terms of Kubernetes batch jobs.

- To create, submit and validate a Kubernetes batch job via HWA.

CHAPTER 10

HWA Integration with Microsoft Azure

In this chapter, we will be focusing about HWA integration with Microsoft Azure. Microsoft Azure provides a range of cloud services, including compute, analytics, storage, and networking. Azure is a cloud computing platform that allows you to access and manage cloud services and resources provided by Microsoft. These services and resources include storing your data and transforming it.

Microsoft Azure job is used to manage the existing VM in Azure subscription as well as creating a new VM. When it comes to managing the VM, we have the following options available:

- Start – start the virtual machine.
- Stop – stop the virtual machine.
- Restart – restart the virtual machine.
- Deallocate – deallocate the virtual machine.
- Generalize – generalize a virtual machine.
- Delete – delete the virtual machine.

Business Requirementax

Let us consider a business scenario, where we want to manage their existing VM. Please see the following configuration steps to manage the VM; in our use case, we are going to start a VM under an Azure subscription.

© Navin Sabharwal and Subramani Kasiviswanathan 2023
N. Sabharwal and S. Kasiviswanathan, *Workload Automation Using HWA*,
https://doi.org/10.1007/978-1-4842-8885-6_10

- We have to create Microsoft Azure job to start a VM. Log in to Dynamic Workload Console and navigate to Design ➤ Manager Workload Definition and select Engine. Refer to Figure 10-1.

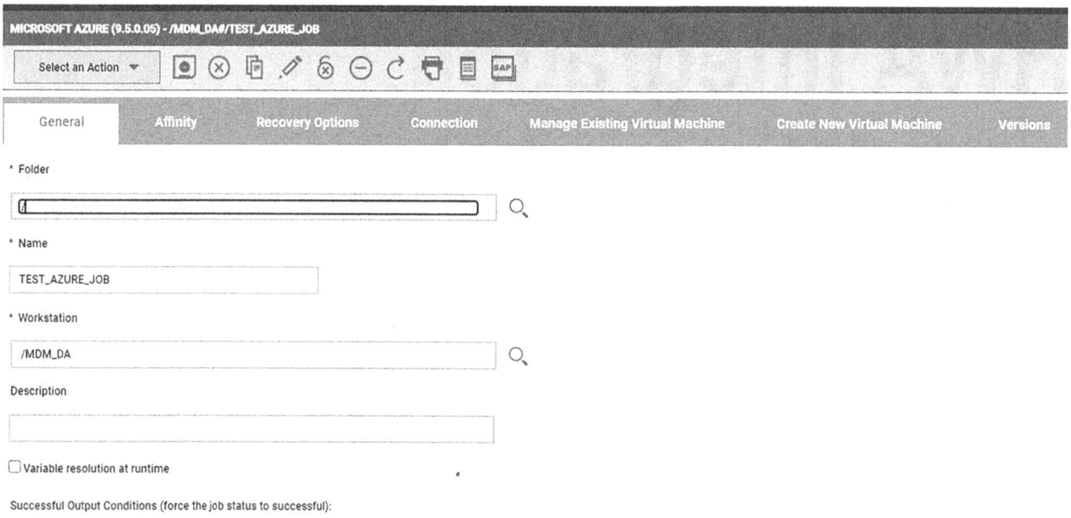
MICROSOFT AZURE (9.5.0.05) - /MDM_DA#/TEST_AZURE_JOB
Select an Action
General
Affinity
Recovery Options
Connection
Manage Existing Virtual Machine
Create New Virtual Machine
Versions
* Folder
* Name
TEST_AZURE_JOB
* Workstation
/MDM_DA
Description
Variable resolution at runtime
Successful Output Conditions (force the job status to successful):

Figure 10-1. *DWC Home Page View*

- This opens the following interface where we have to select job definition and search for Microsoft Azure plug-in. Refer to Figure 10-2.

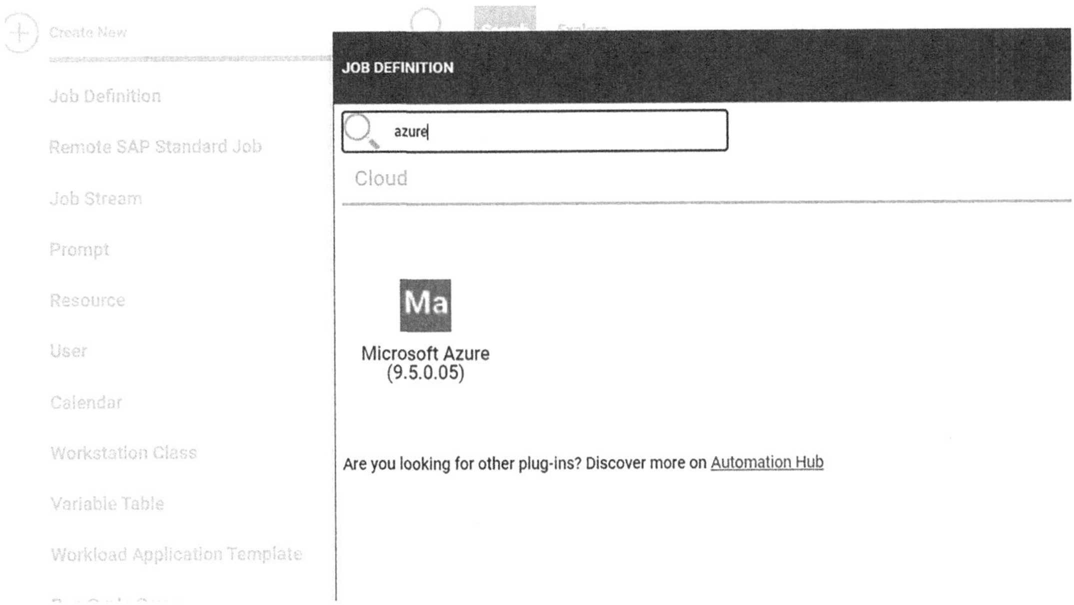

Figure 10-2. *Microsoft Azure Plug-in*

- Let us update the job name and HWA server name as shown in the following. We can give Master dynamic agent to trigger the Microsoft Azure job, and there is no need to install any dynamic agent on customer application server. Refer to Figure 10-3.

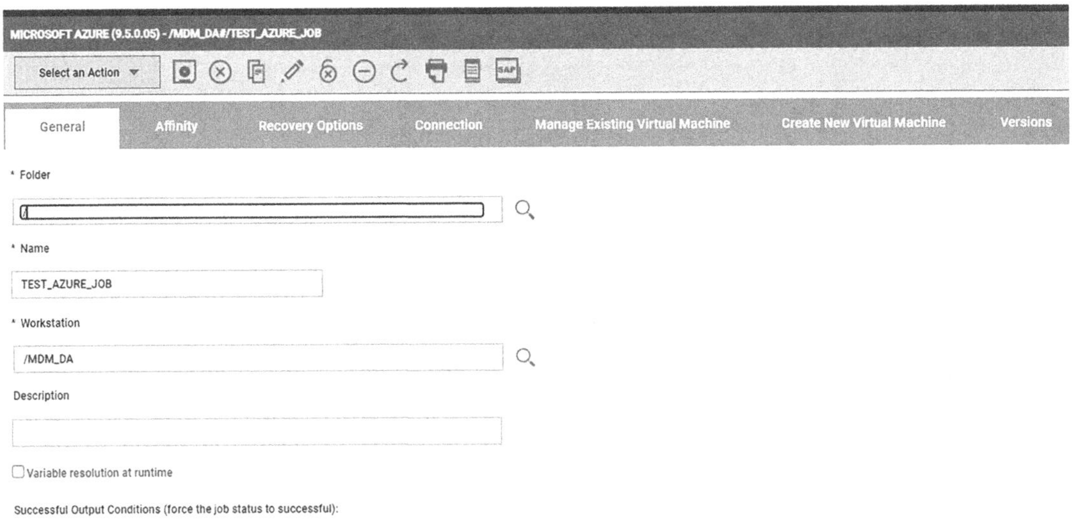

Figure 10-3. *Job Definition Details*

- We need to update subscription, client, tenant, and key on connection tab; please see the following steps to get the connection tab details.

- Login to Azure Portal and select virtual machine to get subscription id. Refer to Figure 10-4.

Figure 10-4. *Subscription Details*

- Select API permissions from Azure Active Directory for Client and Tenant id. Refer to Figure 10-5.

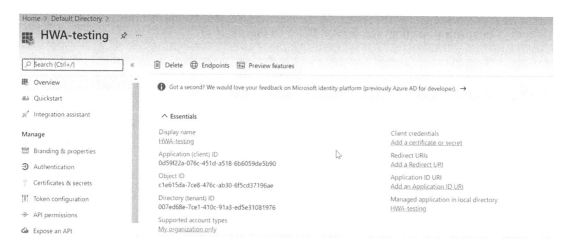

Figure 10-5. *Client and tenant details*

- Select certificates and secrets from Azure Active Directory for key value. Refer to Figure 10-6.

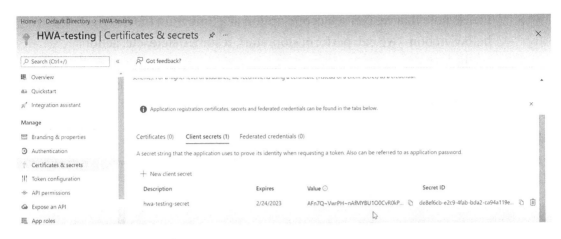

Figure 10-6. *Key Details*

- Use the information gathered from Azure to update the subscription, client, tenant, and key under connection tab in HWA. Refer to Figure 10-7.

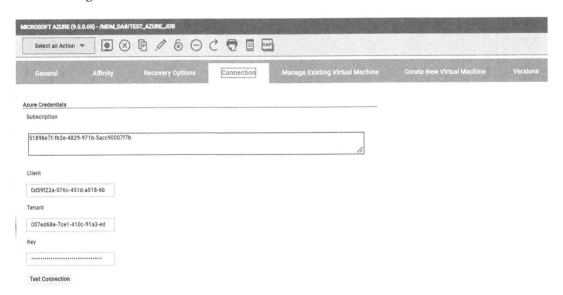

Figure 10-7. *Connection Parameters*

- Make sure test connection is successful to manage the VM.

- Select Manage Existing Virtual Machine tab and update the VM details and select action as "Start." Refer to Figure 10-8.

Please note that we can select one option at a time from the actions listed below. Action – we can perform stop, start, restart, delete, etc. actions on the VM.

Capture a customer image – we can use this option to create the virtual machine image.

Add a tag – select this option to add the name of the tag to the virtual machine.

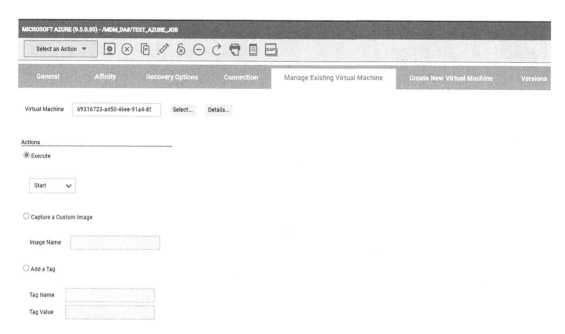

Figure 10-8. *Manage Existing VM Options*

- Please see Figure 10-9 to create a new virtual machine. Please note that we can either manage the VM or create a VM in single job.

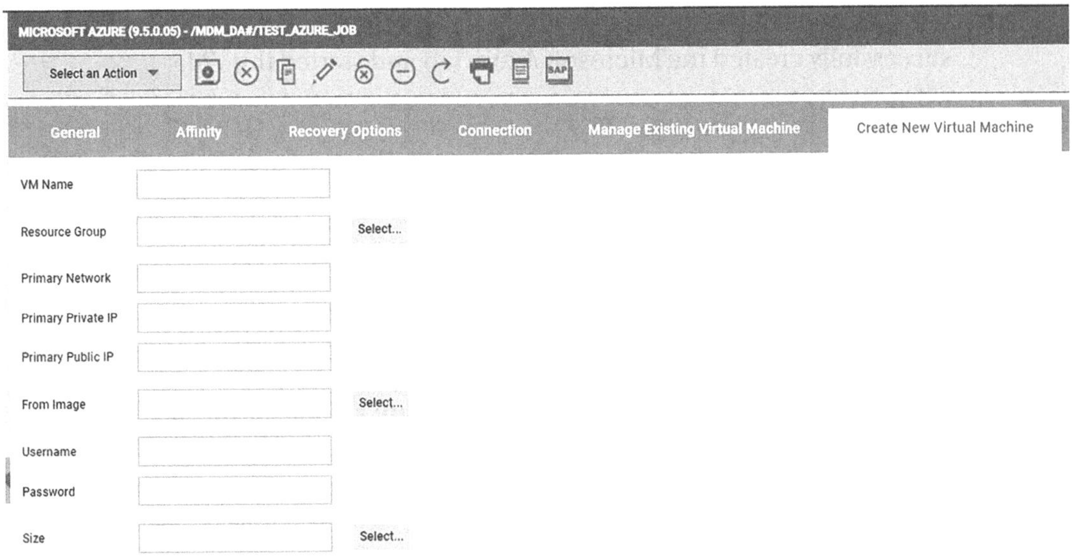

MICROSOFT AZURE (9.5.0.05) - /MDM_DA#/TEST_AZURE_JOB

Select an Action ▾

| General | Affinity | Recovery Options | Connection | Manage Existing Virtual Machine | Create New Virtual Machine |

VM Name

Resource Group   Select...

Primary Network

Primary Private IP

Primary Public IP

From Image   Select...

Username

Password

Size   Select...

Figure 10-9. *Create New VM*

- Save the job and submit the job. To submit the job, navigate to select an action ➤. Submit the job into current plan. Refer to Figure 10-10.

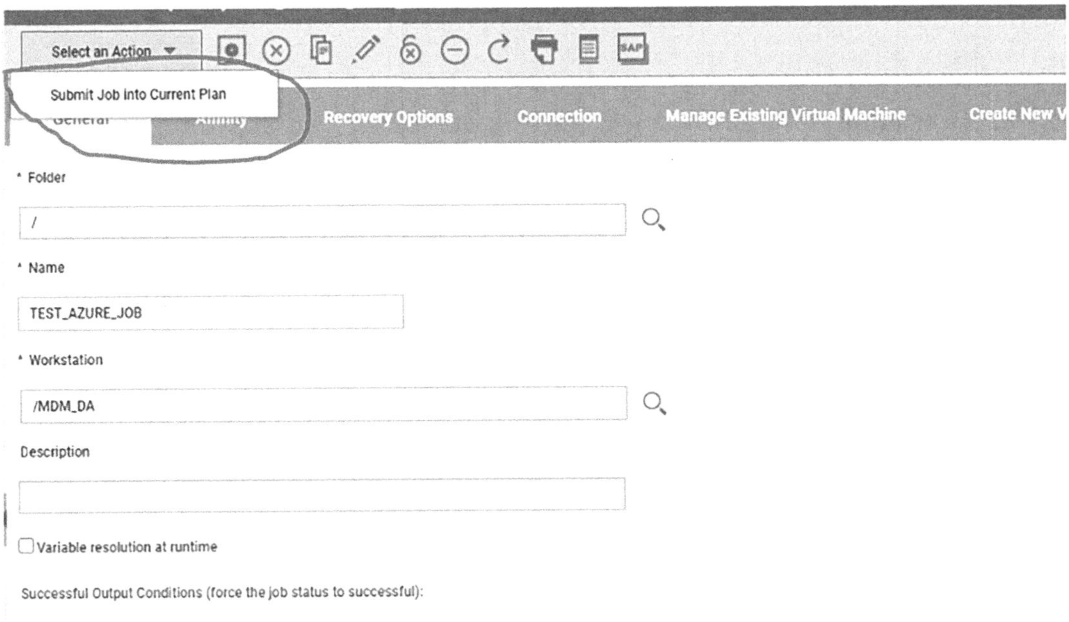

Select an Action ▾

Submit Job into Current Plan

| General | Affinity | Recovery Options | Connection | Manage Existing Virtual Machine | Create New V |

* Folder

/

* Name

TEST_AZURE_JOB

* Workstation

/MDM_DA

Description

☐ Variable resolution at runtime

Successful Output Conditions (force the job status to successful):

Figure 10-10. *Submit Job Action*

- Once the job is successfully executed, validate the log. We have successfully created the Microsoft Azure job and started the VM using HWA.

Summary

In this chapter, we covered the following capabilities of HWA:

- Capabilities of HWA for integration with Microsoft Azure.

- To create, submit and validate a Microsoft Azure job for starting a virtual machine.

CHAPTER 11

HWA Integration with Ansible

In this chapter, we will be covering HWA integration with Ansible. This integration enables Ansible playbooks to be executed on the target machines using HWA.

Ansible can be used to provision the underlying infrastructure of your environment, virtualized hosts and hypervisors, network devices, and bare metal servers. It can also install services and add compute hosts, provision resources, services, and applications inside of your cloud.

To execute Ansible jobs, we require yml and inv files where yml file is the actual execution of the script/command on the target servers and inv file is nothing, the Ansible inventory file defines the hosts and groups of hosts where the commands, modules, and tasks in a playbook execute. Please see the following procedure for creating and executing the Ansible playbooks.

Business Requirement

Let us consider a requirement where we want to execute an Ansible playbook on their application server to process the business data. Let us learn how to configure Ansible job to execute the following requirement.

```
Ansible playbook path    : /SaaS/repo/ans/hwa.yml
Ansible inventory path   : /SaaS/repo/ans/hwa_inv
Ansible/HWA server name  : svcas0166_DA
Ansible job name         : TEST_ANS
```

153

© Navin Sabharwal and Subramani Kasiviswanathan 2023
N. Sabharwal and S. Kasiviswanathan, *Workload Automation Using HWA*,
https://doi.org/10.1007/978-1-4842-8885-6_11

- From the preceding requirement, we have to create HWA Ansible job. To create the job, navigate to Design ➤ Manage Workload Definition ➤ and select the engine. Refer to Figure 11-1.

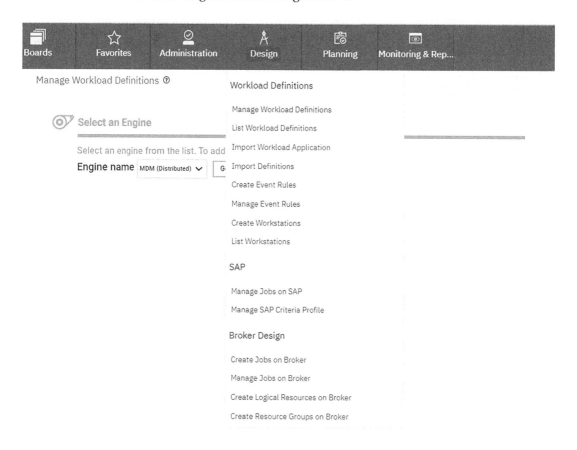

Figure 11-1. *DWC Home Page View*

- This opens the following interface where we have to select job definition and search for Ansible plug-in. Refer to Figure 11-2.

Figure 11-2. *Ansible Plug-in View*

- Please update the job name and HWA server name; in our case, job name is TEST_ANS and server name is svcas0166_DA. Refer to Figure 11-3.

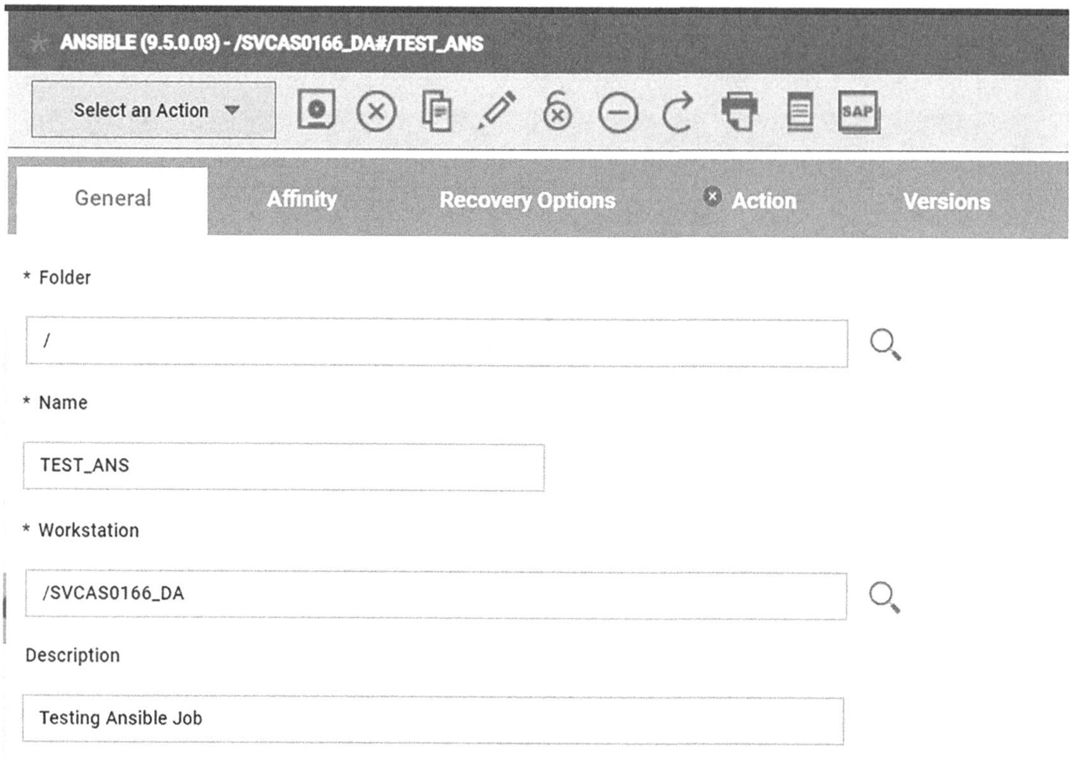

Figure 11-3. *Job Properties*

- Readers need to note that Ansible environment is already installed and configured on the target machines and we are discovering it here and running playbooks from HWA. For creating Ansible playbooks, please refer to Ansible documentation.

- Please note that we can execute Ansible playbooks on HWA dynamic agent, fault tolerant agent, and SSH agent.

  SSH agent is nothing but remote server where HWA application will not be installed on that server but connected to HWA Master/agent server to process the Ansible playbook.

- Select Action tab and update the Playbook and inventory paths. Refer to Figure 11-4 and save the job definition.

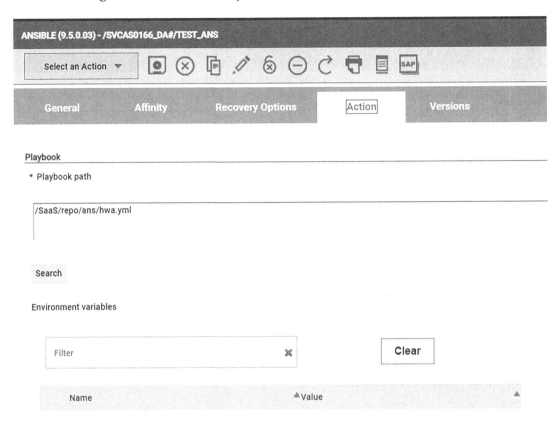

Figure 11-4. *Ansible Action Tab*

- Once the job definition is saved, submit the job to validate the status of the job.

- To submit the job, select an action ➤ Submit job into current plan. Refer to Figure 11-5.

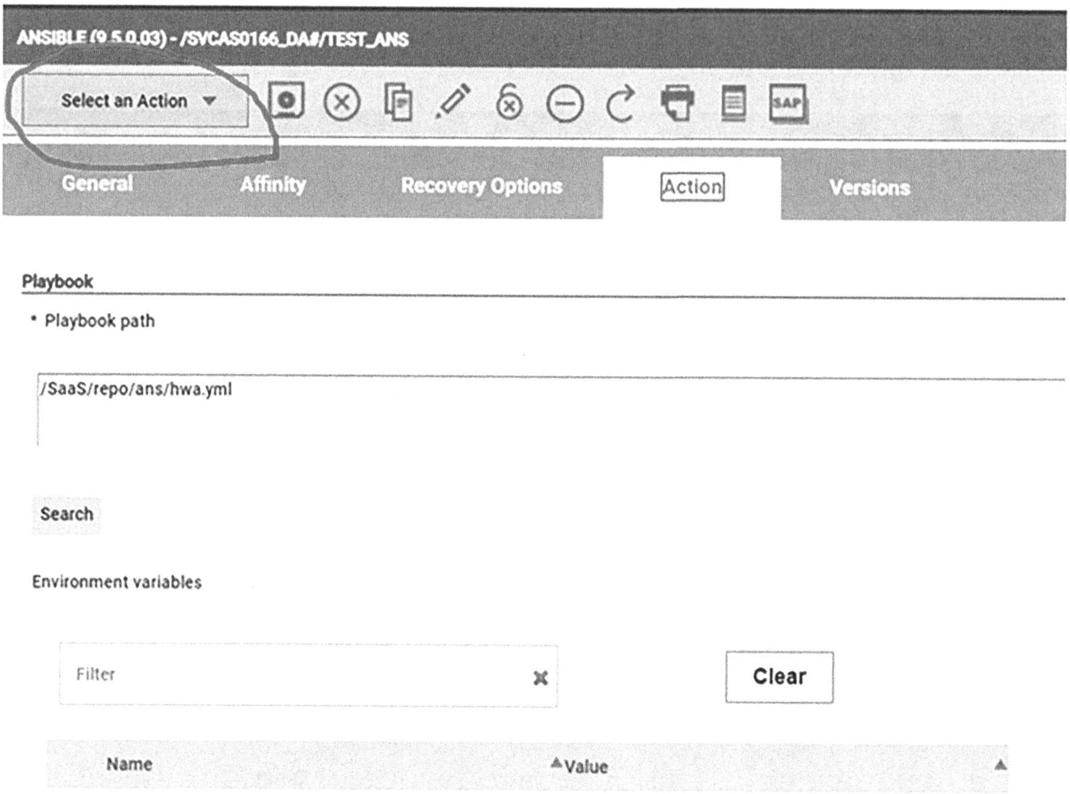

Figure 11-5. *Submit Ansible Job*

- Please see the following job status from the job log for reference.

```
Job log
= Exit Status          : 0
= System time (seconds) : 7    Elapsed time (hh:mm:ss) : 0:00:12
= User time (seconds)   : 6
= Wed 02/23/22 16:31:39 JST
```

- We have successfully created and executed the job.

- We have successfully integrated Ansible job using HWA application.

Summary

In this chapter, we covered the following capabilities of HWA:

- HWA's Ansible plug-in features and options available

- Using HWA to integrate with Ansible and schedule the playbooks, this integration allows to automatically roll out the configurations and deployments to multiple targets.

CHAPTER 12

HWA Event Rule Management

In this chapter, we will cover HWA Event Rule Management. HWA event rule defines a set of actions that are to run upon the occurrence of specific event conditions. The definition of an event rule correlates events and triggers actions. HWA event rules are used to automate the HWA tasks. HWA event rules are created using xml format.

In this chapter, we will discuss about two use cases.

1. HWA integration with ServiceNow for job failure

2. Auto-remediation of HWA process using HWA events

HWA Integration with ServiceNow

ServiceNow is used to highlight and report incidents. Using ServiceNow, we can create the incidents, service requests, and change requests. This use case describes how to integrate HWA with ServiceNow.

When we use HWA in production scenarios, there can be situations where a particular job has failed to execute or has returned with errors. In such scenarios, an incident is generally created in a ticketing system so that administrators can look at the job failure and resolve the incident.

In our use case, we will be creating HWA event to generate job failure ticket in service now. HWA events are used to monitor HWA environment. We can create HWA events to monitor job failures, process down, HWA message files, etc. Based on the event condition, HWA will perform actions like creating a ticket in ServiceNow and sending email to respective DLs. Please see the following procedure to integrate the HWA event with ServiceNow for creating job failure ticket.

© Navin Sabharwal and Subramani Kasiviswanathan 2023
N. Sabharwal and S. Kasiviswanathan, *Workload Automation Using HWA*,
https://doi.org/10.1007/978-1-4842-8885-6_12

We have to create HWA event to generate a job failure ticket in service now. Please see the following steps:

- To create the event rule, navigate to the HWA Dynamic Workload Console and select design. Refer to Figure 12-1.

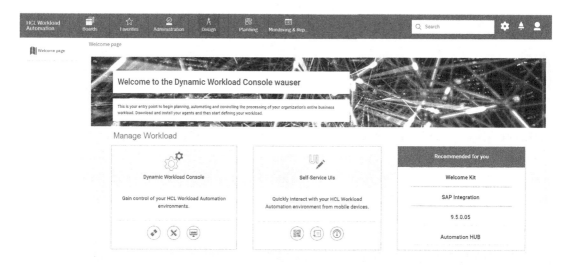

Figure 12-1. *DWC Home Page View*

- From Design tab, navigate to Create Event Rules ➤ Create new ➤ Event rule. Refer to Figure 12-2.

Figure 12-2. *Create New Event Rule*

- This opens the following interface where we need to update the event rule name and select add event. Refer to Figure 12-3.

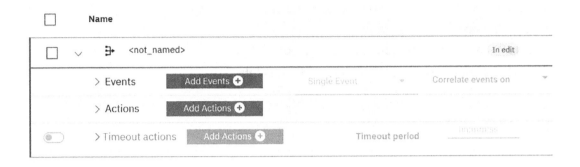

Figure 12-3. *Add Event*

- Select + based on the event rule condition. In our case, we are creating the event rule for job failure. For job failure, select the job status changed option.

Figure 12-4. *Add Event Options*

- This opens the following interface where we need to update the following details. Refer to Figure 12-5.

Job stream workstation - job stream workstation name

Job stream name - job stream name

Job name - name of the job to be monitored

Job workstation - workstation name of the job

Figure 12-5. *Event Parameters*

- We are creating the event for all jobs, so we are updating the value as "*." Please update the status of the job to ERROR status and internal status to ABEND/FAIL to monitor job failure details. Refer to Figure 12-6.

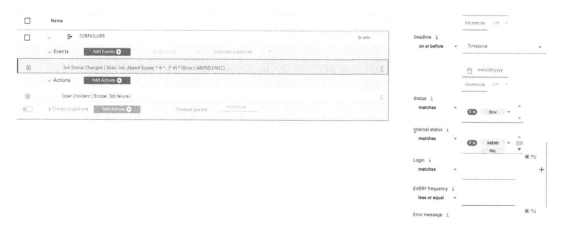

Figure 12-6. *Job Status Options*

- Once event is added, select the Add actions and select the actions based on our requirement. Since we are creating job failure ticket in ServiceNow, we have to select ServiceNow to open incident automatically. Refer to Figure 12-7.

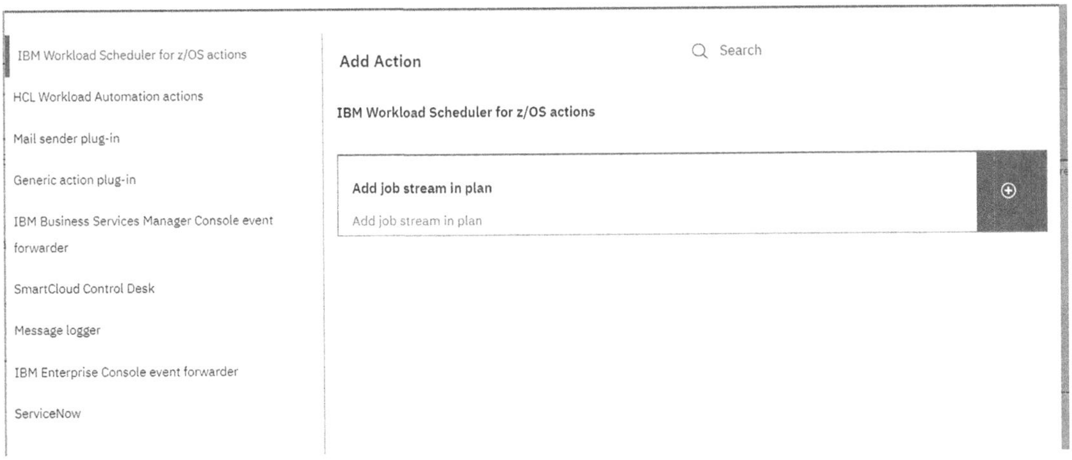

IBM Workload Scheduler for z/OS actions	Add Action	Q Search
HCL Workload Automation actions	IBM Workload Scheduler for z/OS actions	
Mail sender plug-in		
Generic action plug-in	Add job stream in plan	⊕
IBM Business Services Manager Console event forwarder	Add job stream in plan	
SmartCloud Control Desk		
Message logger		
IBM Enterprise Console event forwarder		
ServiceNow		

Figure 12-7. *ServiceNow Action*

- Please fill the required details on the actions added for ServiceNow. Refer to Figure 12-8.

```
Short description :- description of the ticket
ServiceNow URL :- ServiceNow URL
ServiceNow username :- ServiceNow admin username
ServiceNow password :- ServiceNow admin user password
Please add the preceding details in global options as the following:
servicenowUrl / nu =  servicenow URL
servicenowUserName / nn = servicenow username
servicenowUserPassword / np = servicenow password
```

Figure 12-8. *ServiceNow Details*

- Once details are updated, submit the ad hoc job with "exit 1" command and validate if you have received the incident in ServiceNow. To submit the job as ad hoc, refer to Chapter 4.

- We have submitted the job and the job got abend; please see Figure 12-9.

Figure 12-9. *Job Status*

- We can see sample of ServiceNow incident that got created.

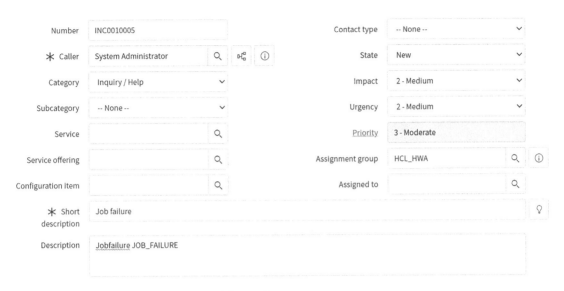

Number	INC0010005			Contact type	-- None --
✳ Caller	System Administrator	🔍 🔀 ⓘ		State	New
Category	Inquiry / Help			Impact	2 - Medium
Subcategory	-- None --			Urgency	2 - Medium
Service	🔍			Priority	3 - Moderate
Service offering	🔍			Assignment group	HCL_HWA 🔍 ⓘ
Configuration item	🔍			Assigned to	🔍
✳ Short description	Job failure				💡
Description	Jobfailure JOB_FAILURE				

Figure 12-10. *ServiceNow Dashboard*

Use Case 2: Auto-Remediation of Process Down

We can also auto-remediate job failures in HWA. In this scenario, rather than opening an incident ticket and waiting for someone to resolve the issue, HWA can auto-remediate some of the issues. We can start HWA process via HWA events automatically. We can monitor the following processes:

- Appservman

- Batchman

- Jobman

- Mailman

- Monman

- Netman

Let us consider a use case where batchman process has to be monitored for Master Domain Manager. Once the batchman process is down, HWA event rule has to start the process automatically to avoid any business impact. Please see the following procedure to create the process down event rule.

- To create the event rule, navigate to design ➤ create new event rule. Select HCL workload automation application monitor ➤ HCL workload automation process not running. Refer to Figure 12-11.

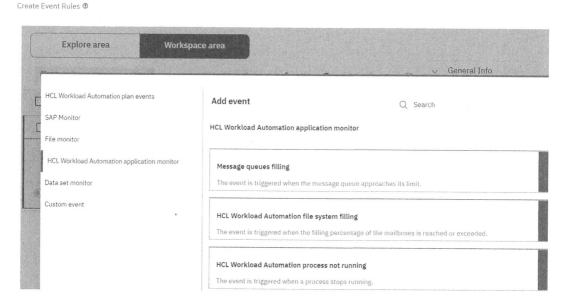

Figure 12-11. *Add Event*

- This opens the following interface where we need to update the event rule name, process name, HCL workload automation installation path, and workstation name. Refer to Figure 12-12.

Figure 12-12. *Event Rule Properties*

Note   Only appserverman, batchman, jobman, mailman, netman, and monman process can be monitored using HWA events, and we can't monitor other application process.

- In our case, we have to select batchman process for MDM. Refer to Figure 12-12.

- Event rule name is AUTO_REMEDIATION. Refer to Figure 12-12.

- Please note that we can give command to start the process or script to start the process. In our case, we are using the script path. Refer to Figure 12-12.
  Please see the following script information:

```
[wauser@svcas0246 logs]$ cat /home/wauser/start.sh
su - wauser -c  ". /HCL/HWA/WA/TWS/StartUp"
su - wauser -c  "cd /HCL/HWA/WA/TWS/bin/ &&  conman start"
su - wauser -c  "cd /HCL/HWA/WA/TWS/bin/ &&  conman startmon"
```

- In action tab, select "run command" and give script path. Refer to Figure 12-13.

Figure 12-13. *Event Details*

- Select the event rule and save. Once the event rule is created, the status of the event rule has to be "active." To check the event rule status, navigate to design ➤ Manage event rule. Refer to Figure 12-14.

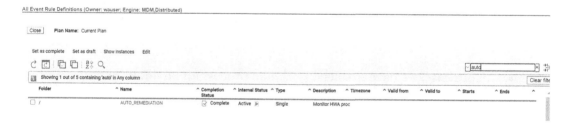

Figure 12-14. *Event Status*

- We can see that event is active. We have successfully created the event rule to monitor the batchman process. Let's login to MDM server and stop the batchman process to validate whether the event rule is working fine.

- Figure 12-15 shows that batchman process is not running.

wauser@svcas0246:/HCL/HWA/WA/TWS

```
[wauser@svcas0246 TWS]$ ps -ef | grep batchman
wauser     4610 18698   0 22:51 pts/4    00:00:00 grep --color=auto batchman
[wauser@svcas0246 TWS]$
```

Figure 12-15. *Batchman Process Is Down*

Log in to Dynamic Workload Console and check the status of MDM
workstation agent process. To check the workstation status, navigate to
Monitoring and Report ➤ Monitor workload ➤ Select Workstation. Refer to
Figure 12-16.

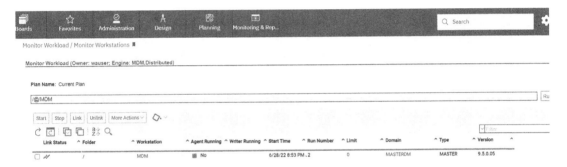

Figure 12-16. *Agent Running Status*

- From Figure 12-16, we can see that agent process is not running; we
  can check the log file <TWS_DATA/stdlist/ appserver/engineServer/
  logs/ messages.log> to validate event rule has started processing.
  Refer to Figure 12-17.

```
by          .
[6/28/22 22:52:09:597 JST] 00000236 com.ibm.tws.cli.model.command.GetModelPropertiesCommand     I AWSJCL050I Command "ls" completed successfully.
[6/28/22 22:52:12:223 JST] 00000028 com.ibm.tws.event.EventProcessorManager                     I AWSEVP001I The following event has been received: event ty
pe = "TWSPROCESSMONITOR"; event provider = "TWSApplicationMonitor"; event scope = "batchman down on MDM".
[6/28/22 22:52:12:231 JST] 00000028 com.ibm.tws.event.plugin.action.ActionHelper                I AWSAHL004I The event rule instance "AUTO_REMEDIATION (6/28
/22 10:52 PM)" has been triggered.
[6/28/22 22:52:12:232 JST] 00000236 com.ibm.tws.event.plugin.action.ActionHelper                I AWSAHL002I The action "RunCommand" for the plug-in "Generi
cActionPlugin" has been started for event rule instance "AUTO_REMEDIATION (6/28/22 10:52 PM)".
[6/28/22 22:52:12:233 JST] 00000028 com.ibm.tws.event.EventProcessorManager                     I AWSEVP007I The following event has matched an existing eve
nt condition: event type = "TWSPROCESSMONITOR"; event provider = "TWSApplicationMonitor"; event scope = "batchman down on MDM".
```

Figure 12-17. *Server Log File*

- From Figure 12-17, we can see that event rule AUTO_REMEDIATION
  started processing and it ran the command to start the batchman

process. Refer to Figure 12-17. Once the event rule action
is performed, we can see that batchman process is started
automatically by the event and agent process is running in
DWC. Refer to Figure 12-18 and 12-19.

```
[wauser@svcas0246 logs]$ ps -ef | grep batchman
wauser    18202 18085  0 22:55 ?       00:00:00 /HCL/HWA/WA/TWS/bin/batchman -parm 32000
wauser    20923 18698  0 22:56 pts/4    00:00:00 grep --color=auto batchman
[wauser@svcas0246 logs]$ █
```

Figure 12-18. *Batchman Process Started*

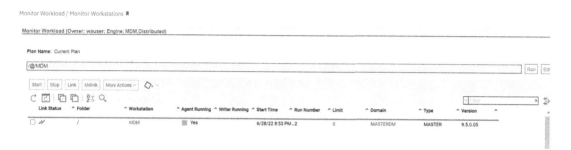

Figure 12-19. *Agent Running Status in Console*

- We have successfully created the process down event rule and started
  the process automatically using HWA events.

Summary

In this chapter, we covered the following capabilities of HWA:

- HWA event rule configuration

- HWA integration with ServiceNow to generate job failure tickets

- Auto-remediation of HWA process down issues via HWA event rule

CHAPTER 13

Tool Administration and Best Practices

In this chapter, we will cover HWA tool administration and best practices. The following topics will be covered as part of this chapter. These topics are required for maintaining the HWA application.

1. Daily application health check

2. Housekeeping procedure to maintain application health

3. Database maintenance procedure

4. Database backup and restore policies

Daily Application Health Check

Daily health check is important for any application. Health check ensures the application is healthy and working as expected. Health check provides the data that is used to anticipate future problems. For HWA application, we should check the following.

- Workstation status – status of the Master Domain Manager and agent servers in console

- Production plan (JnextPlan) – production plan status

- File system and memory usage for Master Domain Manager

The preceding health checks can be automated in HWA using events, and same has been explained in Chapter 12. Alternatively, for monitoring product KPIs, HWA can be integrated with the HERO which is a HWA monitoring and run book automation solution.

© Navin Sabharwal and Subramani Kasiviswanathan 2023
N. Sabharwal and S. Kasiviswanathan, *Workload Automation Using HWA*,
https://doi.org/10.1007/978-1-4842-8885-6_13

To perform the health checks for HWA application, follow the following steps:

Workstation status:

Workstation is nothing but a server where we create and schedule the HWA jobs. If the workstation status is unlinked or process is not running, HWA jobs won't run. To check the status of workstation in HWA, log in to Dynamic Workload Console and navigate to Monitor Workload ➤ Select Workstation ➤ ; refer to Figure 13-1

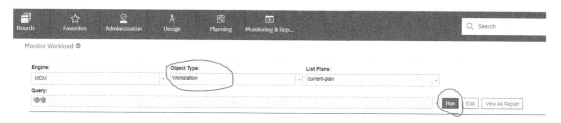

Figure 13-1. *Monitoring Workstation Page*

- This opens the following interface where we can check the status of the workstations. Refer to Figure 13-2.

Monitor Workload / Monitor Workstations

Monitor Workload (Owner: wauser; Engine: MDM,Distributed)

Plan Name: Current Plan

/@/@

| Start | Stop | Link | Unlink | More Actions ∨ | ◇. ∨ |

Link Status	^ Folder	^ Workstation	^ Agent Running	^ Writer Running	^ Start Time	^ Run Number	^ Limit	^ Domain	^ Type	^ Version	^
☐ ⋏ LINKED	/	DWB	Yes	Yes	6/28/22 8:53 PM . 2	0	MASTERDM	Workload Broker	9.5.0.05	
☐ ⋏ LINKED	/	MASTERAGENTS	Yes		6/28/22 8:53 PM . 2	0	MASTERDM	Pool	9.5.0.05	
☐ ⋏	/	MDM	Yes		6/28/22 8:53 PM . 2	0	MASTERDM	MASTER	9.5.0.05	
☐ ⋏ LINKED	/	MDM_DA	Yes		6/28/22 8:53 PM . 2	0	MASTERDM	Agent	9.5.0.05	
☐ ⋏ LINKED	/	MDM_XA	Yes		6/28/22 8:53 PM . 2	0	MASTERDM	Extended Agent	9.5.0.05	
☐ ⋏ LINKED	/	SVALIAPWLAP002	Yes	Yes	6/28/22 5:23 PM . 2	0	MASTERDM	Fault-Tolerant Age	9.5.0.04	

Figure 13-2. *Workstation Status*

- We can see that all the workstations are up and running fine.

Production plan (JnextPlan):

JNextPlan is a script that runs every day in HWA during the start of a defined business day and follows a sequence of steps to move the old plan and generate the new schedule/tasks for the current business day. JnextPlan is created on final job stream. We have to check final job stream status every day. To check the final job stream status, navigate to Monitor workload ➤ Select Jobstream ➤ Final; this will open the following interface to show the status of final job stream. Please refer to Figure 13-3.

Figure 13-3. Final Stream Status

- We can see that final job stream is completed successfully.

  File system and memory:
  HWA Master server is connected to database to create and monitor the jobs across the environment. HWA Master server works as the centralized server where it sends and receives the information across the servers connected to it. We should always maintain HWA Master server health with respect to file system size and memory. To check the file system and memory utilization of the master server, execute "df -h <HWA File system name>" for file system check, and to check the memory, run "free -h"; refer to Figure 13-4.

```
[wauser@SVALIAPWLAP003 wa]$ df -h /HCL/HWA
Filesystem      Size  Used Avail Use% Mounted on
/dev/sdb1        20G  7.5G   13G  38% /HCL/HWA
[wauser@SVALIAPWLAP003 wa]$ free -h
              total        used        free      shared  buff/cache   available
Mem:            15G        7.7G        1.7G        2.3G        6.1G        5.2G
Swap:          2.0G        804M        1.2G
[wauser@SVALIAPWLAP003 wa]$
```

Figure 13-4. File System and Memory Size

- We can see both file system and memory are good, and in case of any issues, please check with system administrator to remove unwanted files.

Housekeeping Procedure to Maintain Application Health

Housekeeping maintains the application healthy and runs the application properly. Housekeeping removes old files or unused files from the application server. For HWA application, we should remove the following log files to maintain the application healthy:

- <HWA_DATA>/stdlist/appserver/engineServer/logs

- <HWA_DATA >/stdlist/logs

- <HWA_DATA >/stdlist/traces

- <HWA_DATA >/schedlog

    Create HWA job to run <HWA_HOME>/TWS/bin/rmstdlist script for removing the logs under <HWA_DATA>/stdlist/. We can remove the logs based on the customer requirement and maintain the required logs for audit purpose. In case if the customer wants to maintain 30 days' logs under <HWA_DATA>/ stdlist/ path, we have to execute the script with "rmstdlist 30" as shown in the following job definition every day. This will keep only 30 days' logs and remove logs which are older than 30 days. We can also create a workstation class for Unix and Windows agents separately to execute the "rmstdlist" script. This script will then get executed on the agents that are a part of the workstation class.

    To create and schedule the job, navigate to Design ->Manage workload definition. Refer to Figure 13-5.

Manage Workload Definitions ⊙

☺ Select an Engine

Select an engine from the list. To add a new engine, browse to Manage engines.

Engine name MDM (Distributed) ∨   Go

Figure 13-5. *DWC Home Page*

- This opens the following interface where we can create the job definition; select job definition and select Unix job as follows .Refer to Figure 13-6.

Figure 13-6. *Unix Job Definition*

- This opens the following interface to create the Unix job. Update job name, server name, and hwa username. The login user has to hwa user since we are removing the hwa logs alone. Refer to Figure 13-7.

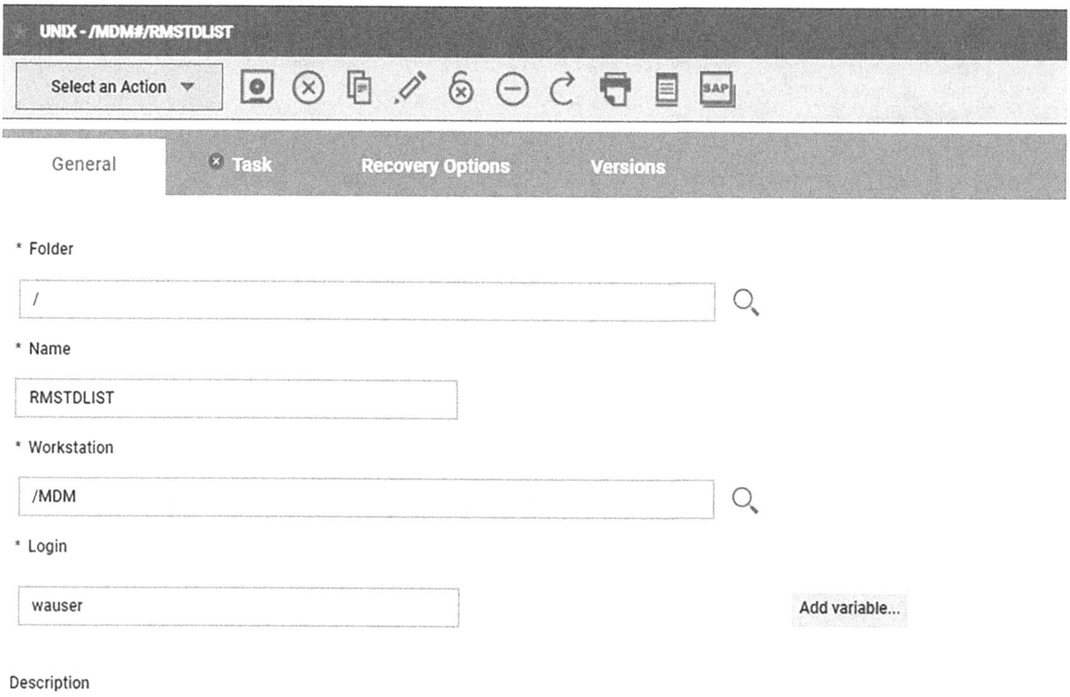

UNIX - /MDM#/RMSTDLIST

Select an Action ▼

General   ⊗ Task   Recovery Options   Versions

* Folder

/

* Name

RMSTDLIST

* Workstation

/MDM

* Login

wauser   Add variable...

Description

Figure 13-7. *HWA Job Properties*

- Under Task tab, update the rmstdlist path to remove logs older than 30 days. In case if we have to remove 90 days' logs, we can change the 30 days to 90. Refer to Figure 13-8.

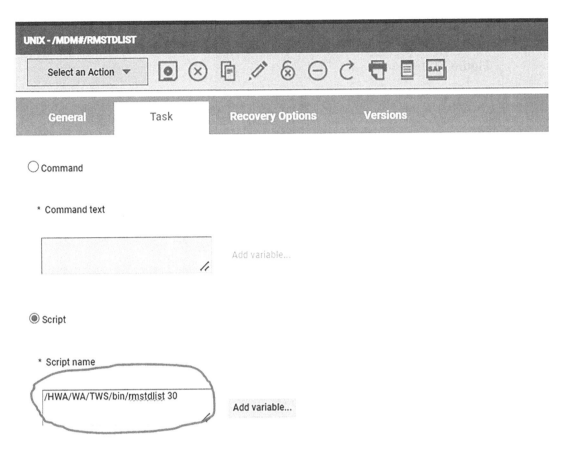

Figure 13-8. *HWA Job Task Properties*

- Select "Save." We have successfully created the job and it will remove the following logs:

```
<HWA_DATA>/stdlist/appserver/engineServer/logs
<HWA_DATA >/stdlist/logs
<HWA_DATA >/stdlist/traces
```

To schedule the job, refer to Chapter 4. To remove the logs under <HWA_DATA >/schedlog path, we have to create a Unix job with command "find /opt/ TWS/twsprd/TWS/schedlog/* -mtime +30 -exec rm {} \;". This command will remove logs older than 30 days. The ideal way to do this is to run the job on both master and backup master; as a best practice, archive the schedlog files on an archiving system for future reference and proceed with removing the logs.

Please see the following procedure to create the job. To create the job, navigate to design ➤ Mange Workload definition ➤ Job definition ➤ Unix job. Refer to Figure 13-9.

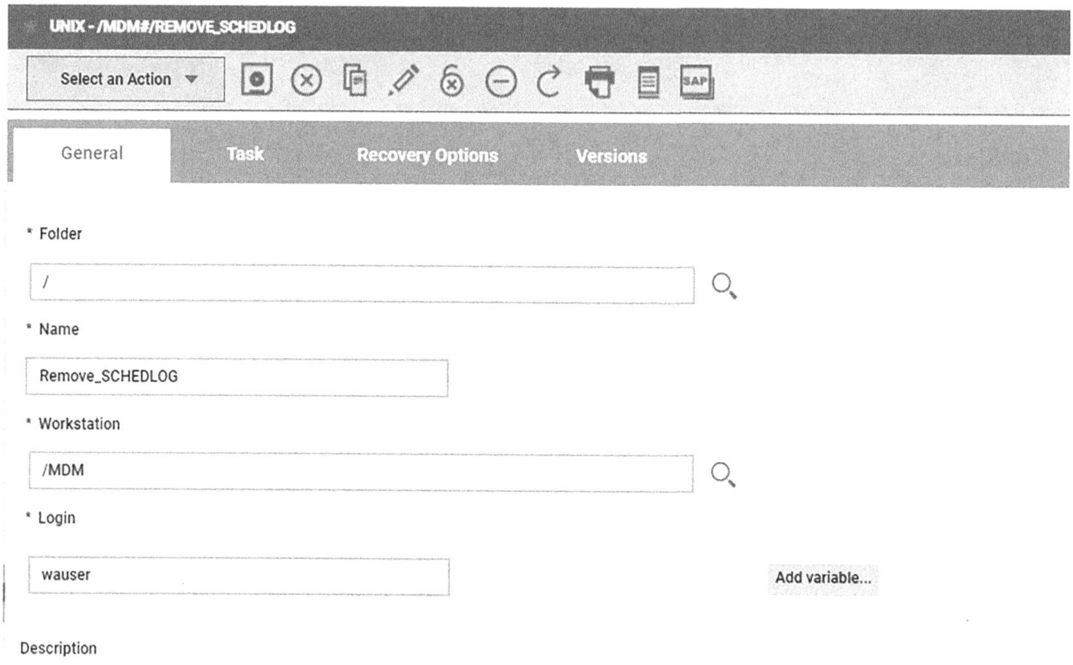

Figure 13-9. *HWA Unix Job Properties*

- Under task tab, in the command section, update "find /opt/TWS/ twsprd/TWS/schedlog/* -mtime +30 -exec rm {} \;" to remove logs older than 30 days. Refer to Figure 13-10.

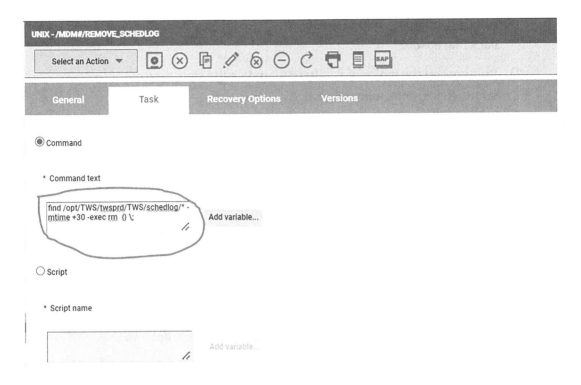

Figure 13-10. *Unix Job Task Properties*

- Select "Save." We have successfully created the job. To schedule the job, refer to Chapter 4. In case if the customer wants to remove the logs older than 90 days, we have to change the command.

```
From
"find /opt/TWS/twsprd/TWS/schedlog/* -mtime +30 -exec rm  {} \;"
To
"find /opt/TWS/twsprd/TWS/schedlog/* -mtime +90 -exec rm  {} \;"
```

Database Maintenance Procedure

Please follow the following high-level steps for database maintenance.

1) Database backup has to be taken as agreed with client. For offline/ online backup procedure, refer to the following link:

www.ibm.com/docs/en/license-metric-tool?topic=database-
backing-up-db2

2) Run stats on the tables of database to update the stats time of
   tables. Refer to the following link:

   www.ibm.com/docs/en/db2/11.5?topic=commands-runstats

3) Reorg on tables of database to reorganize the tables. Downtime as
   per agreed with client. Refer to the following link.

   www.ibm.com/docs/en/db2/11.5?topic=commands-reorg-table

4) Run lower high-water mark on table spaces to reclaim space at
   mount points. Refer to the following link:

   www.ibm.com/support/pages/db2-reducing-size-table-space

5) Auto-maintenance should be set to ON in db cfg in case no
   downtime for reorg on tables, not in case of big tables. Refer to the
   following link.

   www.ibm.com/docs/en/db2/11.5?topic=parameters-auto-
   maint-automatic-maintenance-switches

6) Recreate indexes of tables wherever it is required. Refer to the
   following link.

   www.ibm.com/docs/en/db2-for-zos/11?topic=indexes-types

7) If online backup is configured, include or exclude the archive logs
   in database backup.

8) Archive logs mount point backup to be taken if not included in
   database backup.

9) Pruning of archive logs which have been backed up.
   Points 7, 8, and 9 apply only in case where archival logging is set to
   a path, otherwise, offline backup with downtime in case of circular
   logging.

10) If db2audit log is enabled, mount point housekeeping is required.

    www.ibm.com/docs/en/db2/9.7?topic=commands-db2audit-
    audit-facility-administrator-tool

11) Dialog path needs to be taken care under housekeeping.

```
db2 get dbm cfg |grep -i diag
```

One will get the diagpath and it should be kept below threshold.

Database Backup and Restore Policies

For backup and restore of db2 databases, kindly refer to the following links:

```
www.ibm.com/docs/en/license-metric-tool?topic=database-backing-up-db2
www.ibm.com/docs/en/license-metric-tool?topic=database-restoring-db2
```

Summary

In this chapter, we covered the following capabilities of HWA:

- HWA application health check and housekeeping

- Database backup and restore policies

- Database maintenance

CHAPTER 14

Alerting and Troubleshooting Issues

In this chapter, we are going to discuss HCL workload automation alerts and troubleshooting issues. HWA has a wide variety of alerting mechanism to notify end users to work on their issues on priority. The following three topics describe the alerting and troubleshooting issues in details:

1. Configuring alerts and responses

2. Restoring services post outage

3. Troubleshooting techniques

Configuring Alerts and Responses

HWA alerts are configured using HWA events. Event rules are used to detect event condition and take response actions that need to be performed. An event rule is created in the database. We have explained few use cases in Chapter 12 as well. Let's create a use case for job failure email notification to users.

> Job failure email notification to users:

> Let's create the event rule for job failure email notification; to create the event, navigate to design ➤ create event rule. Refer to Figure 14-1.

© Navin Sabharwal and Subramani Kasiviswanathan 2023
N. Sabharwal and S. Kasiviswanathan, *Workload Automation Using HWA*,
https://doi.org/10.1007/978-1-4842-8885-6_14

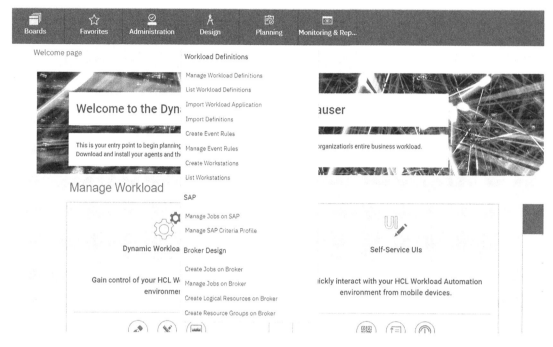

Figure 14-1. *Create Event Rule*

- This opens the following interface where we have to select event rule. Refer to Figure 14-2.

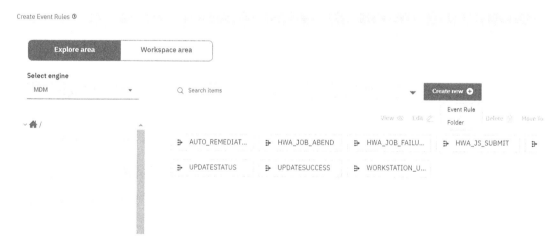

Figure 14-2. *Event Rule Home*

- This opens the following interface where we have to update the event rule name and disable the DRAFT option; if the event rule is created with draft, it will not be activated. Refer to Figure 14-3.

Figure 14-3. *Event Rule General Tab*

- Select Add Events and start creating the event. Refer to Figure 14-4.

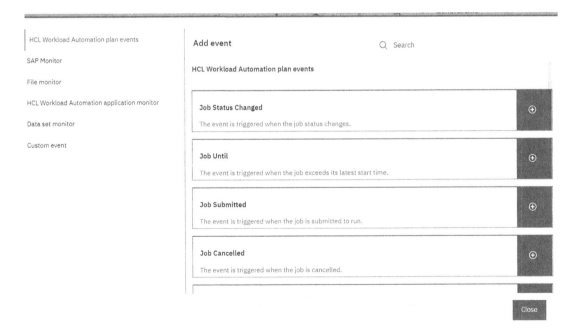

Figure 14-4. *Event Rule Condition Page*

Events are divided into the following major categories:

HCL workload automation object-related events – all the modifications in jobs, job streams, and workstations can be created using these events. Refer to Figure 14-5.

Job	Job Stream	Workstation
Job Status Changed	Job Stream Status Changed	Workstation Status Changed
Job Until	Job Stream Completed	Application Server Status Changed
Job Submitted	Job Stream Until	Child Workstation Link Changed
Job Cancelled	Job Stream Submitted	Parent Workstation Link Changed
Job Restarted	Job Stream Cancelled	
Job Late	Job Stream Late	
Job Promoted		**Prompt**
Job Risk Level Changed		Prompt Status Changed
Job Exceeded Maximum Duration		
Job Did not Reach Minimum Duration		

Figure 14-5. *Event Rule Options*

File monitoring event – events relating to changes to files and logs. Events for files monitoring that are predefined are Log message written, File created, File deleted, and Modification completed.

Application monitoring event – events relating to HWA processes, file system, and message box. Events for Application monitoring that are predefined are Message queues filling, HCL Workload Automation file system filling, and HCL Workload Automation process not running.

SAP-related event – these events are available only if you have installed HWA for applications. Using these events, we can detect if a specified string is matched in the log file or when the specified IDoc is created.

- In our case, select job status changed and select "+" symbol; refer to Figure 14-6.

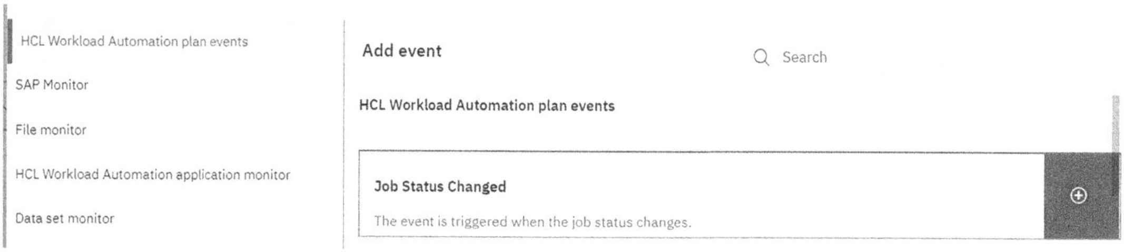

Figure 14-6. *Job Status Change Condition*

- This opens the following interface where we need to update the following details. Refer to Figure 14-7.

```
Job stream workstation - job stream workstation name
Job stream name - job stream name
Job name - name of the job to be monitored
Job workstation - workstation name of the job
```

Figure 14-7. *Event Rule Properties*

- We are creating the event for all jobs, so we are updating the value as "*." Please update the status of the job to ERROR status and internal status to ABEND/FAIL to monitor job failure details. Refer to Figure 14-8.

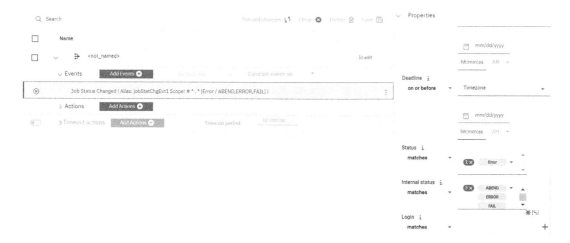

Figure 14-8. *Event Rule Properties*

- Once event is added, select the Add actions "+" and select the actions based on our requirement. Since we are creating job failure email notification, we have to select mail sender plug-in and send mail to send emails for job failures. Refer to Figure 14-9.

Figure 14-9. *Event Rule Action Properties*

- We have to update recipient address, subject, and body of the mail as in Figure 14-9. And select "SAVE." In HWA, there are multiple options

like including HWA variables that contain information about the
HWA objects in subject and body of the mail as shown in Figure 14-9.

- We have successfully created email notification for job failure event,
  and please see the following figure for sample email notification for
  job failure. Refer to Figure 14-10.

From: TWS@svcas0246.localdomain <TWS@svcas0246.localdomain>
Sent: Friday, July 8, 2022 1:18 AM
Subject: HWA Job HWA_TEST_ABEND abended

HWA Job HWA_TEST_ABEND abended in workstation MDM

Figure 14-10. *Sample Email Notification*

Follow the preceding steps to create similar event rule for other workload scheduler
objects.

Restore Services Post Outage

Outage is nothing but services that are not available. There are different types of outages.

1. Planned outage

2. Unplanned outage

3. Emergency outage

During planned outages, we can stop the applications and inform the application
status to respective team to proceed further with the outage. To stop the HWA
application during planned outages, log in to HWA master server and execute the
following commands from the following paths to stop the Liberty process. Refer to
Figure 14-11.

```
<HWA_HOME/appservertools/stopAppServer.sh>
<DWC_HOME/appservertools/stopAppServer.sh>
```

```
[root@svcas0246 appservertools]# ls
appserverstart.sh  setEnv.sh  startAppServer.sh  stopAppServer.sh
[root@svcas0246 appservertools]# ./stopAppServer.sh
Setting CLI environment variables....
HCL Workload Scheduler Environment Successfully Set !!!
HCL Workload Automation Environment Successfully Set !!!
HCL Workload Automation(UNIX)/CONMAN 9.5.0.05 (20211214)
Licensed Materials - Property of IBM* and HCL**
5698-WSH
(C) Copyright IBM Corp. 1998, 2016 All rights reserved.
(C) Copyright HCL Technologies Ltd. 2016, 2021 All rights reserved.
* Trademark of International Business Machines
** Trademark of HCL Technologies Limited
Installed for user "wauser".
Locale LANG set to the following: "en"
Scheduled for (Exp) 07/07/22 (#3) on MDM.  Batchman LIVES.  Limit: 0, Fence: 0, Audit Level: 1
/
%stopappserver; wait
AWSBHU622I A stop command was issued for the application server on workstation "MDM".
[root@svcas0246 appservertools]# cd /HCL/HWA/DWC/appservertools
[root@svcas0246 appservertools]# ./stopAppServer.sh

Stopping server dwcServer.
Server dwcServer stopped.
[root@svcas0246 appservertools]#
```

Figure 14-11. *Liberty Process Stop Command*

- Check the process is stopped by executing "ps -ef | grep java." Refer to Figure 14-12.

```
[wauser@svcas0246 appservertools]$ ps -ef | grep java
root       1589     1  0 May24 ?        00:46:41 /opt/vmware-jre/bin/java -jar /opt/vmware-appdirector/agent/nobel-agent.jar
wauser    23139 13106  0 00:06 pts/0    00:00:00 grep --color=auto java
[wauser@svcas0246 appservertools]$
```

Figure 14-12. *Liberty Process Status*

- We can see that Liberty process stopped; we have to stop the other HWA process. To stop other HWA process, execute the following commands. Refer to Figure 14-13.

```
conman stop
conman shut;wait
conman stopmon
ShutDownLwa
```

---

wauser@svcas0246:/HCL/HWA/DWC/appservertools

```
[wauser@svcas0246 appservertools]$ conman stop
HCL Workload Automation(UNIX)/CONMAN 9.5.0.05 (20211214)
Licensed Materials - Property of IBM* and HCL**
5698-WSH
(C) Copyright IBM Corp. 1998, 2016 All rights reserved.
(C) Copyright HCL Technologies Ltd. 2016, 2021 All rights reserved.
* Trademark of International Business Machines
** Trademark of HCL Technologies Limited
Installed for user "wauser".
Locale LANG set to the following: "en"
Scheduled for (Exp) 07/07/22 (#3) on MDM.  Batchman down.  Limit: 0, Fence: 0, Audit Level: 1
/
%stop
AWSBHU013I MDM already stopped.
[wauser@svcas0246 appservertools]$ conman shut;wait
HCL Workload Automation(UNIX)/CONMAN 9.5.0.05 (20211214)
Licensed Materials - Property of IBM* and HCL**
5698-WSH
(C) Copyright IBM Corp. 1998, 2016 All rights reserved.
(C) Copyright HCL Technologies Ltd. 2016, 2021 All rights reserved.
* Trademark of International Business Machines
** Trademark of HCL Technologies Limited
Installed for user "wauser".
Locale LANG set to the following: "en"
Scheduled for (Exp) 07/07/22 (#3) on MDM.  Batchman down.  Limit: 0, Fence: 0, Audit Level: 1
/
%shut
[wauser@svcas0246 appservertools]$ conman stopmon
HCL Workload Automation(UNIX)/CONMAN 9.5.0.05 (20211214)
Licensed Materials - Property of IBM* and HCL**
5698-WSH
(C) Copyright IBM Corp. 1998, 2016 All rights reserved.
(C) Copyright HCL Technologies Ltd. 2016, 2021 All rights reserved.
* Trademark of International Business Machines
** Trademark of HCL Technologies Limited
Installed for user "wauser".
Locale LANG set to the following: "en"
Scheduled for (Exp) 07/07/22 (#3) on MDM.  Batchman down.  Limit: 0, Fence: 0, Audit Level: 1
/
%stopmon
The command was forwarded to monman for MDM
[wauser@svcas0246 appservertools]$ ShutDownLwa
Stopping tebctl-tws_cpa_agent_wauser Agent (32011):...........................OK
[wauser@svcas0246 appservertools]$ █
```

Figure 14-13. *HWA Process Stop Commands*

- To check the process are stopped, please execute "ps -ef | grep TWS." Refer to Figure 14-14.

```
Stopping tebctl-tws_cpa_agent_wauser Agent (32011):...........................OK
[wauser@svcas0246 appservertools]$ ps -ef | grep TWS
wauser    3274 13106  0 00:10 pts/0    00:00:00 grep --color=auto TWS
[wauser@svcas0246 appservertools]$ █
```

Figure 14-14. *HWA Process Status Command*

- We can see that all the HWA processes are stopped successfully. In case of any issues while starting the application, check the log files <HWA_DATE/stdlist> for reference.

Whereas in unplanned or emergency outages we will not be having enough notification to stop the application, all of a sudden our application will stop without intimation. During this time, we should check with our respective platform team (OS team) about the details. Once the outage is completed, we should start the application from scratch. To start the HWA application from scratch, follow the following steps.

- Log in to the Master server and execute "StartUp" command. Refer to Figure 14-15.

```
wauser@svcas0246:/HCL/HWA/WA/TWS

  login as: wauser
  wauser@svcas0246.dryicelabs.com's password:
Last login: Thu Jul  7 22:43:31 2022 from 172.16.1.50
Setting CLI environment variables....
HCL Workload Scheduler Environment Successfully Set !!!
HCL Workload Automation Environment Successfully Set !!!
HCL Workload Scheduler Environment Successfully Set !!!
[wauser@svcas0246 TWS]$ StartUp
TWS for UNIX/STARTUP
Licensed Materials - Property of IBM* and HCL**
5698-WSH
(C) Copyright IBM Corp. 1998, 2016 All rights reserved.
(C) Copyright HCL Technologies Ltd. 2016, 2021 All rights reserved.
* Trademark of International Business Machines
** Trademark of HCL Technologies Limited
HCL Workload Scheduler Environment Successfully Set !!!
Operation successful.
Program code level: 20211214
Setting CLI environment variables....
HCL Workload Scheduler Environment Successfully Set !!!
HCL Workload Automation Environment Successfully Set !!!
HCL Workload Scheduler Environment Successfully Set !!!
HCL Workload Automation(UNIX)/CONMAN 9.5.0.05 (20211214)
Licensed Materials - Property of IBM* and HCL**
5698-WSH
(C) Copyright IBM Corp. 1998, 2016 All rights reserved.
(C) Copyright HCL Technologies Ltd. 2016, 2021 All rights reserved.
* Trademark of International Business Machines
** Trademark of HCL Technologies Limited
Installed for user "wauser".
Locale LANG set to the following: "en"
Scheduled for (Exp) 07/07/22 (#3) on MDM.  Batchman LIVES.  Limit: 0, Fence: 0, Audit Level: 1
/
%startappserver; wait
AWSBHU620I A start command was issued for the application server on workstation "MDM".
[wauser@svcas0246 TWS]$
```

Figure 14-15. *HWA Process Start Command*

- Once StartUp Command is issued, check the processes that are started by using "ps -ef | grep TWS." Refer to Figure 14-16.

```
[wauser@svcas0246 TWS]$ ps -ef | grep TWS
wauser    3694 16271  0 23:06 pts/7    00:00:00 grep --color=auto TWS
wauser   14624     1  0 Jun28 ?        00:00:06 /HCL/HWA/WA/TWS/bin/netman
wauser   25258 14624  0 23:03 ?        00:00:00 /HCL/HWA/WA/TWS/bin/appservman -- 2021 MDM CONMAN UNIX 9.5.0.05 MESSAGE
wauser   25529     1 99 23:03 ?        00:02:56 /HCL/HWA/WA/TWS/JavaExt/jre/jre/bin/java -javaagent:/HCL/liberty/wlp/wlp/bin/tools/ws-javaagent.jar -Djava.aw
t.headless=true -Djdk.attach.allowAttachSelf=true -Djava.library.path=/usr/Tivoli/TWS/GSKit64/8/lib64/:/usr/Tivoli/TWS/OpenSSL64/1.1/lib64:/HCL/HWA/WA/TWS/bi
n/:/HCL/HWA/WA/TWS/jdbcdrivers/db2: -Dorg.eclipse.emf.ecore.EPackage.Registry.INSTANCE=com.ibm.sdo.jsdl.helper.SDORegistry -DITDWB_HOME= /HCL/HWA/WA/TDWB -Dc
om.ibm.correlation.expressions.java.ExpressionLanguage.classpath=/HCL/HWA/WA/usr/servers/engineServer/apps/TWSEngineModel.ear/*: -Djava.security.properties=/
HCL/HWA/WA/usr/servers/engineServer/wa.java.security -Xms2048m -Xmx4096m -Xgcpolicy:gencon -jar /HCL/liberty/wlp/wlp/bin/tools/ws-server.jar engineServer
wauser   32011     1  0 Jun27 ?        00:00:02 /HCL/HWA/WA/TWS/ITA/cpa/ita/agent
wauser   32030 32011  0 Jun27 ?        00:09:07 /HCL/HWA/WA/TWS/bin/JobManager
[wauser@svcas0246 TWS]$
```

Figure 14-16. *HWA Process Status*

- Once the process is started using StartUp, we should check database connectivity using "optman ls" command; in case of any issues, check <HWA_DATA/stdlist/> log files; the output of the command should be like in Figure 14-17.

```
[wauser@svcas0246 TWS]$ optman ls
HCL Workload Automation(UNIX)/OPTMAN 9.5.0.05 (20211214)
Licensed Materials - Property of IBM* and HCL**
5698-WSH
(C) Copyright IBM Corp. 1998, 2016 All rights reserved.
(C) Copyright HCL Technologies Ltd. 2016, 2021 All rights reserved.
* Trademark of International Business Machines
** Trademark of HCL Technologies Limited
Installed for user "wauser".
Locale LANG set to the following: "en"

TECServerName / th = localhost
TECServerPort / tp = 5529
approachingLateOffset / al = 120
auditHistory / ah = 400
auditStore / as = BOTH
baseRecPrompt / bp = 1000
bindUser / bu = wauser
carryStates / cs =
companyName / cn = HCL
deadlineOffset / do = 2
defaultWksLicenseType / wn = PERSERVER
deploymentFrequency / df = 5
```

Figure 14-17. *Db Connectivity Check*

- We can see that database is accessible from HWA; we can start the remaining process. Please follow the following steps to start other process like batchman, monman, mailman, etc. Execute the following commands; refer to Figure 14-18.

conman start

conman startmon

StartUpLwa

```
[wauser@svcas0246 TWS]$ conman start
HCL Workload Automation(UNIX)/CONMAN 9.5.0.05 (20211214)
Licensed Materials - Property of IBM* and HCL**
5698-WSH
(C) Copyright IBM Corp. 1998, 2016 All rights reserved.
(C) Copyright HCL Technologies Ltd. 2016, 2021 All rights reserved.
* Trademark of International Business Machines
** Trademark of HCL Technologies Limited
Installed for user "wauser".
Locale LANG set to the following: "en"
Scheduled for (Exp) 07/07/22 (#3) on MDM.  Batchman LIVES.  Limit: 0, Fence: 0, Audit Level: 1
/
%start
AWSBHU014I MDM already active.
[wauser@svcas0246 TWS]$ conman startmon
HCL Workload Automation(UNIX)/CONMAN 9.5.0.05 (20211214)
Licensed Materials - Property of IBM* and HCL**
5698-WSH
(C) Copyright IBM Corp. 1998, 2016 All rights reserved.
(C) Copyright HCL Technologies Ltd. 2016, 2021 All rights reserved.
* Trademark of International Business Machines
** Trademark of HCL Technologies Limited
Installed for user "wauser".
Locale LANG set to the following: "en"
Scheduled for (Exp) 07/07/22 (#3) on MDM.  Batchman LIVES.  Limit: 0, Fence: 0, Audit Level: 1
/
%startmon
AWSBHU450I Monitoring is already active for MDM.
[wauser@svcas0246 TWS]$ StartUpLwa
tebctl-tws_cpa_agent_wauser already started
```

Figure 14-18. *HWA Start Commands*

- Once the commands are executed, check the process status by executing "ps -ef | grep TWS." In case of any issues, check <HWA_DATA/stdlist/> log files. Refer to Figure 14-19.

```
[wauser@svcas0246 TWS]$ ps -ef | grep TWS
wauser   12545 14624  0 23:09 ?        00:00:00 /HCL/HWA/WA/TWS/bin/mailman -parm 32000 -- 2002 MDM CONMAN UNIX 9.5.0.05 MESSAGE
wauser   12546 14624  0 23:09 ?        00:00:00 /HCL/HWA/WA/TWS/bin/monman -- 2011 MDM CONMAN UNIX 9.5.0.05 MESSAGE
wauser   12718 14624  0 23:09 ?        00:00:00 /HCL/HWA/WA/TWS/bin/writer -- 2001 DWB MAILMAN UNIX 9.5.0.05 13
root     12789     1  0 23:09 ?        00:00:00 ./ssmagent.bin -f /HCL/HWA/WA/TWSDATA/ssm/config
wauser   12926 12545  0 23:09 ?        00:00:00 /HCL/HWA/WA/TWS/bin/batchman -parm 32000
root     12990 12926  0 23:09 ?        00:00:00 /HCL/HWA/WA/TWS/bin/jobman
wauser   14624     1  0 Jun28 ?        00:00:06 /HCL/HWA/WA/TWS/bin/netman
wauser   17208 16271  0 23:11 pts/7    00:00:00 grep --color=auto TWS
wauser   25258 14624  0 23:03 ?        00:00:00 /HCL/HWA/WA/TWS/bin/appservman -- 2021 MDM CONMAN UNIX 9.5.0.05 MESSAGE
wauser   25529     1 43 23:03 ?        00:03:17 /HCL/HWA/WA/TWS/JavaExt/jre/jre/bin/java -javaagent:/HCL/liberty/wlp/wlp/bin/tools/ws-javaagent.jar -Djava.a▸
t.headless=true -Djdk.attach.allowAttachSelf=true -Djava.library.path=/usr/Tivoli/TWS/GSKit64/8/lib64/:/usr/Tivoli/TWS/OpenSSL64/1.1/lib64:/HCL/HWA/WA/TWS/bi
n/:/HCL/HWA/WA/TWS/jdbcdrivers/db2: -Dorg.eclipse.emf.ecore.EPackage.Registry.INSTANCE=com.ibm.sdo.jsdl.helper.SDORegistry -DITDWB_HOME= /HCL/HWA/WA/TDWB -Dc
om.ibm.correlation.expressions.java.ExpressionLanguage.classpath=/HCL/HWA/WA/usr/servers/engineServer/apps/TWSEngineModel.ear/*: -Djava.security.properties=/
HCL/HWA/WA/usr/servers/engineServer/wa.java.security -Xms2048m -Xmx4096m -Xgcpolicy:gencon -jar /HCL/liberty/wlp/wlp/bin/tools/ws-server.jar engineServer
wauser   32011     1  0 Jun27 ?        00:00:02 /HCL/HWA/WA/TWS/ITA/cpa/ita/agent
wauser   32030 32011  0 Jun27 ?        00:09:07 /HCL/HWA/WA/TWS/bin/JobManager
[wauser@svcas0246 TWS]$
```

Figure 14-19. *HWA Process Status*

- We can see that all the HWA processes are started; now change directory to <DWC_HOME/appservertools> and execute startAppServer.sh script to start the DWC Liberty process. Refer to Figure 14-20.

```
[root@svcas0246 appservertools]# ./startAppServer.sh

Starting server dwcServer.
Server dwcServer started with process ID 25419.
[root@svcas0246 appservertools]# pwd
/HCL/HWA/DWC/appservertools
[root@svcas0246 appservertools]#
```

Figure 14-20. *DWC Process Start Command*

- Once the process is started, check the status of process by executing "ps -ef | grep DWC." In case of any issues, check <DWC_DATA/ stdlist/> log files. Refer to Figure 14-21.

```
[root@svcas0246 appservertools]# ./startAppServer.sh

Starting server dwcServer.
Server dwcServer started with process ID 25419.
[root@svcas0246 appservertools]# pwd
/HCL/HWA/DWC/appservertools
[root@svcas0246 appservertools]# ps -ef | grep DWC
root     25419     1 99 23:13 pts/9    00:01:52 /HCL/HWA/DWC/java/jre/bin/java -javaagent:/HCL/liberty/wlp/wlp/bin/tools/ws-javaagent.jar -Djava.awt.headless
=true -Djdk.attach.allowAttachSelf=true -Djava.library.path=/HCL/HWA/DWC/jdbcdrivers/db2: -Djava.awt.headless=true -Djava.security.properties=/HCL/HWA/DWC/us
r/servers/dwcServer/wa.java.security -Xms2048m -Xmx4096m -Xgcpolicy:gencon -jar /HCL/liberty/wlp/wlp/bin/tools/ws-server.jar dwcServer
root     30491  1099  0 23:15 pts/9    00:00:00 grep --color=auto DWC
[root@svcas0246 appservertools]#
```

Figure 14-21. *DWC Liberty Process Status*

- We can see that DWC Liberty process is started. Log in to the DWC console to validate DWC is working fine. Refer to Figure 14-22.

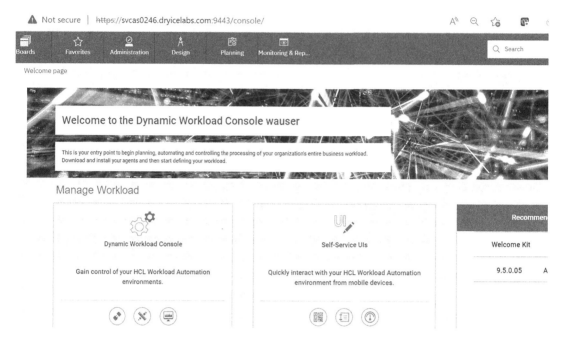

Figure 14-22. *DWC Home Page*

Troubleshooting Techniques

In this part, we are going to discuss about HWA application troubleshooting steps and its fixes. Please see the following use cases:

- HWA agent process down.

- Workstation unlinked status.

- Dynamic workload console is not reachable.

HWA agent process down:

Let us consider SVALIAPWLAP002 agent process is down. To troubleshoot the issue, first log in to Dynamic Workload Console and navigate to Monitoring and Reporting ➤ Monitor workload ➤ select workstation and run. Refer to Figure 14-23.

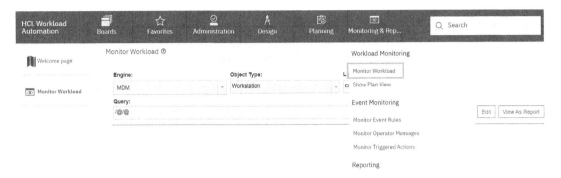

Figure 14-23. *Monitor Workload Page*

- This opens the following interface where we can check the status of the agents. Refer to Figure 14-24.

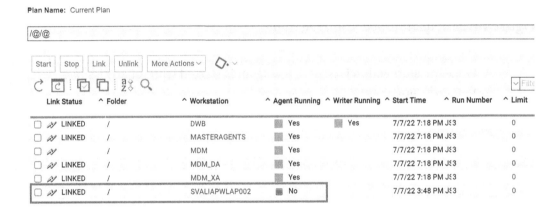

Figure 14-24. *Agent Status Page*

- We can see that agent process is down for SVALIAPWLAP002 server. To start the agent, select the agent SVALIAPWLAP002 check box and select start option to start the process. Refer to Figure 14-25.

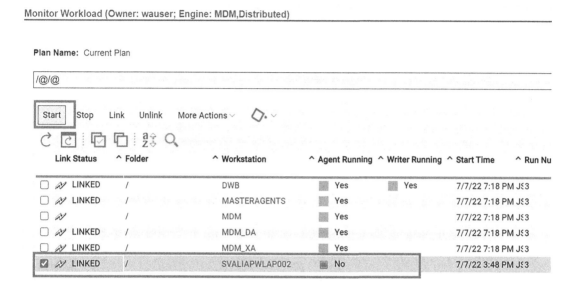

Figure 14-25. *Agent Process Start Action*

- Once the start action is issued, the agent is started without any issues. Refer to Figure 14-26.

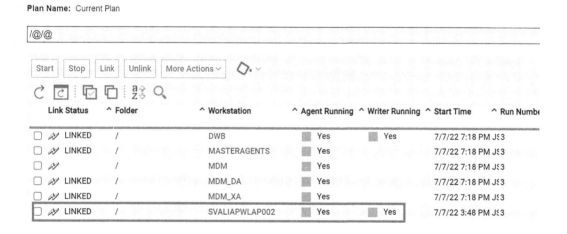

Figure 14-26. *Agent Status*

- In case of any issues, log in to the agent server and check the logs under <HWA_DATA/stdlist/traces>.

Workstation unlinked status:

Let us consider SVALIAPWLAP002 is unlinked, and we got the alert for the same. To troubleshoot the issue, first log in to Dynamic Workload Console and navigate to Monitoring and Reporting ➤ Monitor workload ➤ select workstation and run. Refer to Figure 14-27.

Figure 14-27. *Monitor Workload*

- This opens the following interface where we can check the status of the workstation; refer to Figure 14-28.

Plan Name: Current Plan

/@/@

| Start | Stop | Link | Unlink | More Actions ∨ |

	Link Status	^ Folder	^ Workstation	^ Agent Running	^ Writer Running	^ Start Time	^ Run Number	^ Lin
☐	LINKED	/	DWB	Yes	Yes	7/7/22 7:18 PM JS3	0	
☐	LINKED	/	MASTERAGENTS	Yes		7/7/22 7:18 PM JS3	0	
☐		/	MDM	Yes		7/7/22 7:18 PM JS3	0	
☐	LINKED	/	MDM_DA	Yes		7/7/22 7:18 PM JS3	0	
☐	LINKED	/	MDM_XA	Yes		7/7/22 7:18 PM JS3	0	
☐	UNLINKED	/	SVALIAPWLAP002	Yes		7/7/22 3:48 PM JS3	0	

Figure 14-28. *Agent Status*

- We can see that the agent status is unlinked; select the agent check box and select link option to link the agent server. Refer to Figure 14-29.

Plan Name: Current Plan

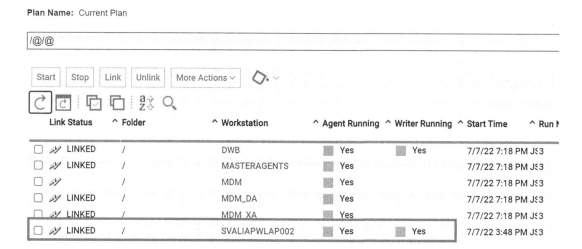

Figure 14-29. *Agent Is Linked*

- This will link the agent to Master server; even after providing the link if the agent is still unlinked, check the agent connectivity to Master server by running telnet command as the following. Refer to Figure 14-30.

```
[hwauser@SVALIAPWLAP002 ~]$ telnet svcas0246 31111
Trying 10.1.150.90...
Connected to svcas0246.
Escape character is '^]'.
Connection closed by foreign host.
[hwauser@SVALIAPWLAP002 ~]$
```

Figure 14-30. *telnet Connection*

Note   svcas0246 is the master server.

- If the issue is still existing, check the agent and master log files. Path for both the files are <HWA_DATA/stdlist/traces>.

Dynamic Workload Console is not reachable:

Let us consider DWC is not reachable and we got the mail from monitoring team. First check DWC access by logging into DWC link (`https://host/ ip:9443/console`); if dwc is not accessible, then there may be an issue with DWC Liberty process. Log in to Master server and switch to <DWC/ appservertools> as the following and execute startAppServer.sh to start the DWC Liberty process. Refer to Figure 14-31 to start the DWC Liberty process.

```
/ꞏꞏꞏꞏ/ꞏꞏꞏꞏ/ꞏꞏꞏ/ꞏꞏꞏꞏꞏꞏꞏꞏꞏꞏꞏꞏ
[wauser@svcas0246 appservertools]$ ls -ltr
total 24
-rwxr-xr-x. 1 root root 9510 Jun 20 21:37 appserverstart.sh
-rwxr-xr-x. 1 root root 2034 Jun 20 21:37 stopAppServer.sh
-rwxr-xr-x. 1 root root  832 Jun 20 21:38 setEnv.sh
-rwxr-xr-x. 1 root root 3171 Jun 20 21:38 startAppServer.sh
[wauser@svcas0246 appservertools]$ ./startAppServer.sh
```

Figure 14-31. *DWC Liberty Start*

- Validate whether DWC is accessible by logging into the DWC console. If the liberty process is up and still it is not accessible, check the logs under <DWC_DATA/stdlist>.

Summary

In this chapter, we covered the following capabilities of HWA:

- HWA alerts and configuration.

- HWA troubleshooting steps for few use cases.

- HWA stop and start steps in detail.

CHAPTER 15

HWA Reporting

In this chapter, we will focus on HWA reporting features. HWA has a built-in reporting feature to generate, validate, and compare the job runtime, status, executions, and statistics. There are two types of reporting available in HWA:

1. Predefined reports

2. Personalized reports

Predefined reports have a defined category in which we can select our desired data and provide the required information and run the report. In predefined report, we can generate different types of reports like the following

- Job Run Statistics – to check successful jobs %, errors %, job late start time, long running job statistics, etc.

- Job Run History – to calculate how many times a particular application job has run over a period.

- Workstation Workload Summary – we can check how much workload a job is utilizing on that server.

- Workstation Workload Runtimes – workstation runtime details.

- Actual Production Details – information to be displayed in the actual production details report output.

- Planned Production Details – you can specify time intervals and have only the jobs that ran in these time frames included in the report output.

- Custom SQL Report – we can run a database query by giving the required table details.

© Navin Sabharwal and Subramani Kasiviswanathan 2023
N. Sabharwal and S. Kasiviswanathan, *Workload Automation Using HWA*,
https://doi.org/10.1007/978-1-4842-8885-6_15

Let's generate the predefined report for makeplan job in DWC; to generate the report, log in to DWC and navigate to Monitoring and Reporting ➤ Manage Predefined reports. Refer to Figure 15-1

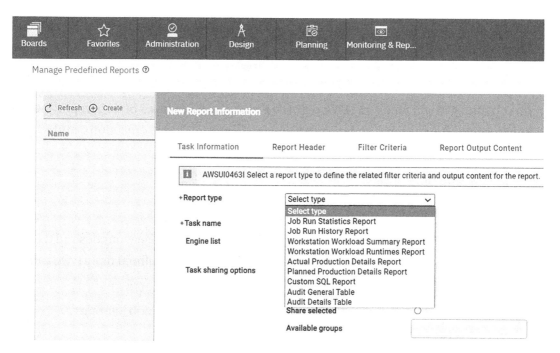

Figure 15-1. *Reporting Page*

- Select "Job Run History Report" and select Report Header tab.

- Update the required description of the report. Refer to Figure 15-2

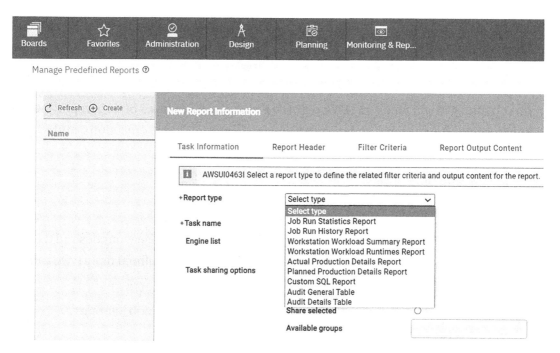

Figure 15-2. *Report Header*

- Select Filter Criteria tab and update workstation name, job stream, and job details, and specify from and to dates. Refer to Figure 15-3

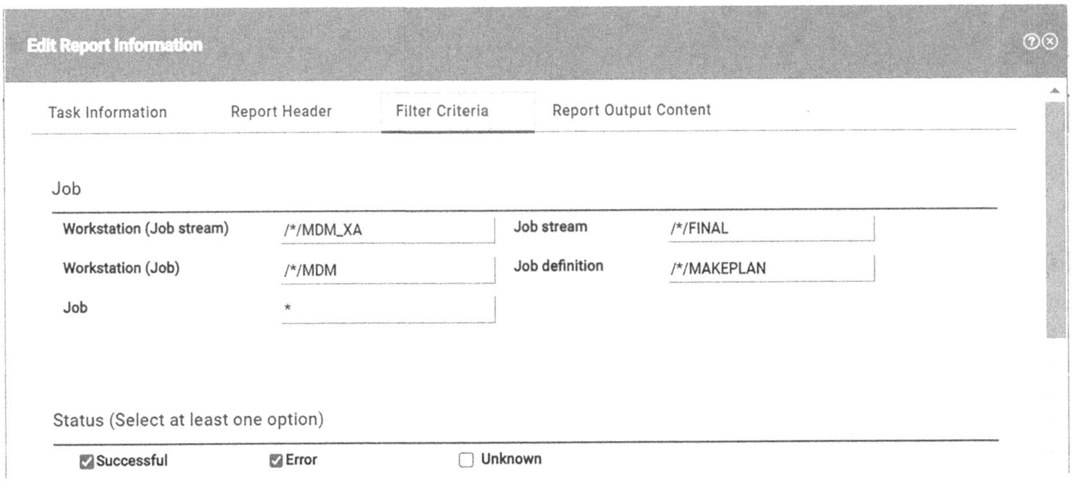

Figure 15-3. *Filter Criteria*

- Select the output report format; in HWA, we have HTML, PDF, and CSV formats available. Refer to Figure 15-4. In our case, we are generating PDF format.

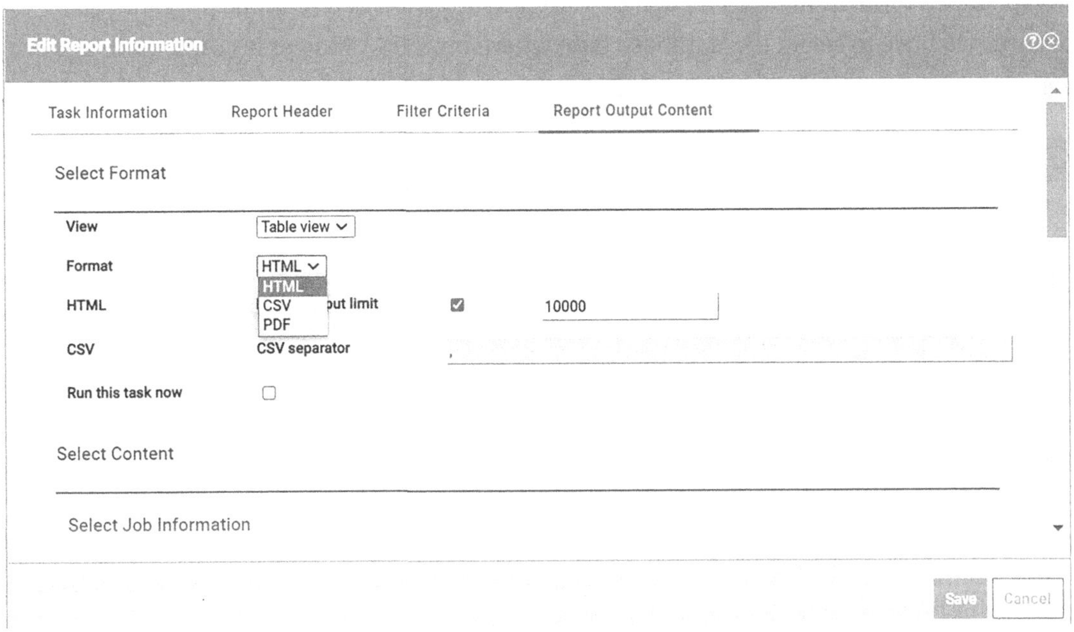

Figure 15-4. *Output Format*

- Once all the details are provided and saved, we can see the report name as the following. Refer to Figure 15-5. Now we can run the report using Run option as shown in Figure 15-5.

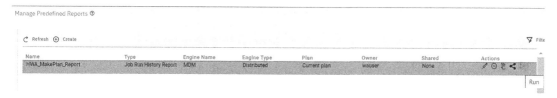

Figure 15-5. *Report Name Page*

- Once the report is executed, we will get the output in PDF format. Refer to Figure 15-6.

Figure 15-6. *Output Report Format*

In personalized reports, the administrator can manage and import personalized reports created with BIRT (Business Intelligence and Reporting Tool) and execute it. To generate the personalized reports, navigate to Monitoring and Reporting ➤ Manage Personalized reports. Refer to Figure 15-7. In personalized and manage personalized reports, we have to select the database type where our HWA Master is connected. In our case, database type is db2 and we are generating report for all jobs.

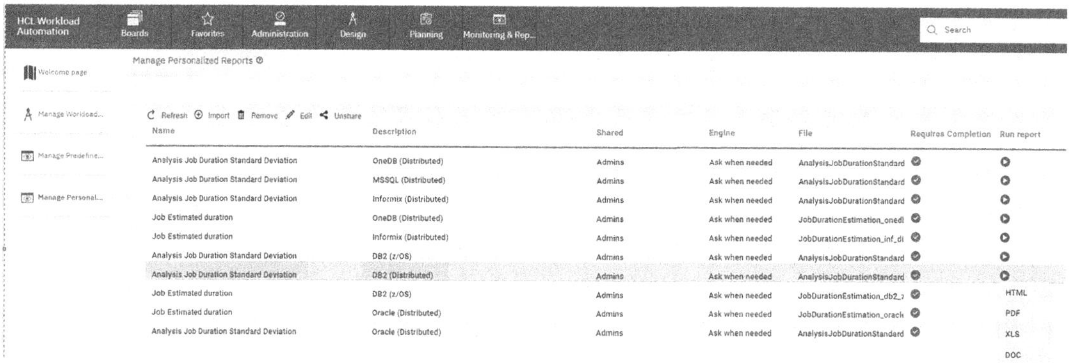

Figure 15-7. *Managed Personalized Report Page*

- This opens the following interface where we need to give the job name, job stream, and workstation. Since we are generating report for all jobs, select "*." Refer to Figure 15-8.

Figure 15-8. *Managed Personalized Report Parameters*

- Select "run" to generate the report. Refer to Figure 15-9.

Figure 15-9. *Report Output*

- We can see that report has been generated successfully; perform the same steps for personalized reports to get the required outputs.

Note   In Hwa, we have some custom BIRT report sample for HWA available on `https://github.com/WorkloadAutomation/ReportSamples` for reference.

Summary

In this chapter, we covered the following capabilities of HWA:

- HWA reporting capabilities and its features
- How to create HWA reports and download them for reference

CHAPTER 16

HWA Security

In this chapter, we will discuss HWA security. HWA security is managed through the security file. The purpose of the security file is to control access on one or more HWA objects created in database and plan. The security file information consists of user and groups with permissions to perform certain tasks. In HWA, we can give access to the users using two models:

1. Traditional model (file based)

2. Role-based access control

Traditional Model (File Based)

This option is disabled after the master domain manager installation by default. To enable the traditional model, set "enRoleBasedSecurityFileCreation" to "Yes." In this model, we can access security file using "dumpsec" and "makesec" commands from the command-line interface. Dumpsec command is used to decrypt the current security file into an editable configuration, whereas makesec command is used to encrypt the security file and apply the modifications. We should add the users in HWA Master server as well as DWC to give respective roles and access for users. The HWA Dynamic Workload Console has the following roles to give access to users:

Administrator – the administrator can access the security, the workload definitions, workload submission, forecast, SAP, and the section monitoring and reporting.

User management – the admin can customize the access on the environment.

© Navin Sabharwal and Subramani Kasiviswanathan 2023
N. Sabharwal and S. Kasiviswanathan, *Workload Automation Using HWA*,
https://doi.org/10.1007/978-1-4842-8885-6_16

Operator – the operator can manage the available plans and create a trial plan and a forecast plan. In addition, it can access workload and event monitoring.

Developer – the developer can access the workload definitions and can view the preproduction plan.

Analyst – the analyst can only manage the workload reports.

Reporting – the operator can generate the reports from the preproduction plan.

View access – only display and list access for objects.

Please follow the following procedure to create scheduling and operator users in DWC. The username details are as follows:

Scheduling user : HWA_Sched

Operator user : HWA_Operator

To create the new users, navigate to Administration tab ➤ Manage Roles.

Figure 16-1. *Manage Roles*

- This opens the interface shown in Figure 16-2, and as per our requirement, add the "HWA_Sched" user under Developer role.

Figure 16-2. *Selection of Roles*

- To add the user, select Entities option as shown in Figure 16-2.

- This opens the interface as in Figure 16-3; select "Add."

Figure 16-3. *Selection of Users*

- This opens the interface as in Figure 16-4 and add the user HWA_
  Sched and select "Add entity" to add the user under Developer role
  and click "save."

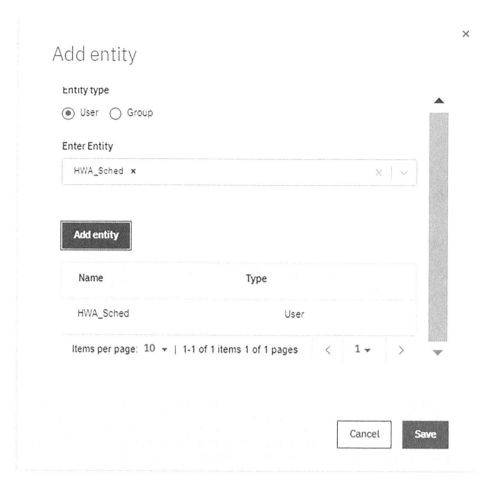

Figure 16-4. *Addition of Users into Developer Role*

- We can see that user HWA_Sched has been successfully added to the
  Developer role as shown in Figure 16-5.

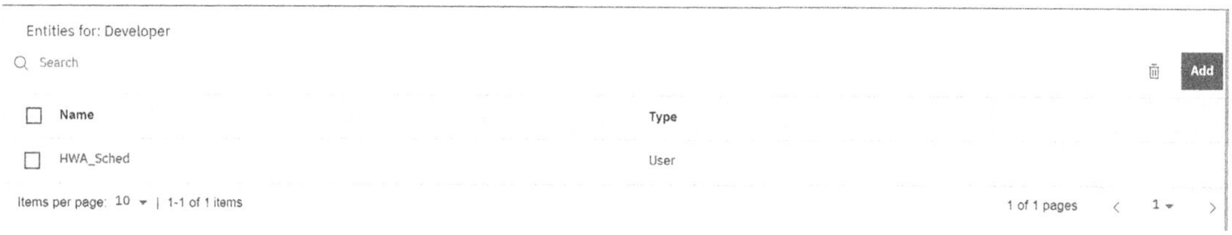

Entities for: Developer

Q Search

	Name	Type
☐	HWA_Sched	User

Items per page: 10 ▾ | 1-1 of 1 items   1 of 1 pages   <   1 ▾   >

Figure 16-5. *User Added to Developer Role View*

- Perform the preceding same steps as to HWA_Opertor but add the user under "Operator" role as shown in Figure 16-6.

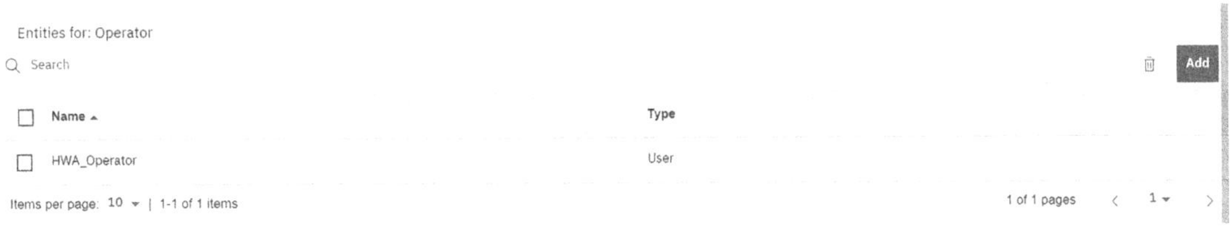

Entities for: Operator

Q Search   🗑   Add

☐   Name ▲   Type

☐   HWA_Operator   User

Items per page: 10 ▾ | 1-1 of 1 items   1 of 1 pages   ⟨   1 ▾   ⟩

Figure 16-6.   *User Added to Developer Role View*

Once we have added the roles to the user (local user), we must add the users in authentication_config.xml files under HWA and DWC paths, path for the files as follows:

```
<Inst_dir>/usr/servers/engineServer/configDropins/overrides/
authentication_config.xml
<Inst_dir>/usr/servers/dwcServer/configDropins/overrides/
authentication_config.xml
```

- Please refer to Figure 16-7 to update the users HWA_Sched and HWA_Operator on authentication_config.xml file.

```
[wauser@SVALIAPWLAP002 overrides]$ cat authentication_config.xml
<server description="basicRealm">

        <basicRegistry id="basic" realm="TWSRealm">
                    <!--
                            This user is defined in wauser_variables.xml,
                            and it is the user used by liberty to run, if you remove
                            this user please set another valid user and password
                            defined into the user registry in wauser_variables.xml.
                -->
            <user name="${user.twsuser.id}" password="${user.twsuser.password}"/>
            <user name="HWA_Operator" password="HWA_Operator"/>
            <user name="HWA_Sched" password="HWA_Sched"/>
        </basicRegistry>
</server>
[wauser@SVALIAPWLAP002 overrides]$ compass
```

Figure 16-7.   *Authentication File*

- Once the users are added in xml files, let's export the existing security file information; to export the existing information, execute dumpsec command as the following. Refer to Figure 16-8.

```
[wauser@svcas0246 TWS]$ dumpsec >access.txt
HCL Workload Scheduler (UNIX)/DUMPSEC 9.5.0.05 (20211214)
Licensed Materials - Property of IBM* and HCL**
5698-WSH
(C) Copyright IBM Corp. 1998, 2016 All rights reserved.
(C) Copyright HCL Technologies Ltd. 2016 All rights reserved.
* Trademark of International Business Machines
** Trademark HCL Technologies Limited
[wauser@svcas0246 TWS]$ ▓
```

Figure 16-8. *dumpsec Command*

- Modify access.txt file and add the users "HWA_Sched and HWA_ operator" like the following. Refer to Figure 16-9 and save the file.

```
USER FULLCONTROLofALLOBJECTS
        CPU=@+LOGON=root,wauser,HWA_Sched,HWA_operator▓
BEGIN
        USEROBJ CPU=@    ACCESS=ADD,DELETE,DISPLAY,MODIFY,USE,ALTPASS,LIST,UNLOCK
        JOB     CPU=@+NAME=@    ACCESS=ADD,ADDDEP,ALTPRI,CANCEL,CONFIRM,DELDEP,DELETE,DISPLAY,KILL,MODIFY,RELEASE,REPLY,RERUN,SUBMIT,USE,LIST,UNLOCK,SUBMITDB
, RUN
        SCHEDULE        CPU=@+NAME=@    ACCESS=ADD,ADDDEP,ALTPRI,CANCEL,DELDEP,DELETE,DISPLAY,LIMIT,MODIFY,RELEASE,REPLY,SUBMIT,LIST,UNLOCK
        RESOURCE        CPU=@+NAME=@    ACCESS=ADD,DELETE,DISPLAY,MODIFY,RESOURCE,USE,LIST,UNLOCK
        PROMPT  NAME=@  ACCESS=ADD,DELETE,DISPLAY,MODIFY,REPLY,USE,LIST,UNLOCK
        FILE    NAME=@  ACCESS=BUILD,DELETE,DISPLAY,MODIFY,UNLOCK
        CPU     CPU=@   ACCESS=ADD,CONSOLE,DELETE,DISPLAY,FENCE,LIMIT,LINK,MODIFY,SHUTDOWN,START,STOP,UNLINK,USE,LIST,UNLOCK,RUN,RESETFTA,MANAGE
        PARAMETER       CPU=@+NAME=@    ACCESS=ADD,DELETE,DISPLAY,MODIFY,LIST,UNLOCK
        CALENDAR        NAME=@  ACCESS=ADD,DELETE,DISPLAY,MODIFY,USE,LIST,UNLOCK
        REPORT  NAME=@  ACCESS=DISPLAY
        EVENTRULE       NAME=@  ACCESS=ADD,DELETE,DISPLAY,MODIFY,LIST,UNLOCK
        ACTION  PROVIDER=@      ACCESS=DISPLAY,SUBMIT,USE,LIST
        EVENT   PROVIDER=@      ACCESS=USE
        VARTABLE        NAME=@  ACCESS=ADD,DELETE,DISPLAY,MODIFY,USE,LIST,UNLOCK
        WKLDAPPL        NAME=@  ACCESS=ADD,DELETE,DISPLAY,MODIFY,LIST,UNLOCK
        RUNCYGRP        NAME=@  ACCESS=ADD,DELETE,DISPLAY,MODIFY,USE,LIST,UNLOCK
        LOB     NAME=@  ACCESS=USE
CONTINUE
```

Figure 16-9. *Modifying the File*

- Once the file is saved, execute makesec command as the following. Refer to Figure 16-10 to give access to the user.

```
[wauser@svcas0246 TWS]$ makesec access.txt
HCL Workload Scheduler (UNIX)/MAKESEC 9.5.0.05 (20211214)
Licensed Materials - Property of IBM* and HCL**
5698-WSH
(C) Copyright IBM Corp. 1998, 2016 All rights reserved.
(C) Copyright HCL Technologies Ltd. 2016 All rights reserved.
* Trademark of International Business Machines
** Trademark HCL Technologies Limited
AWSDEK313W You have role-based security enabled. If you generate the
security file on a master domain manager or backup master domain manager,
the security file will be overwritten by an automated procedure at any
time.
MAKESEC:Starting user FULLCONTROLOFALLFOLDERS [access.txt (#2)]
MAKESEC:Starting user FULLCONTROLOFALLOBJECTS [access.txt (#18)]
MAKESEC:Done with access.txt, 0 errors (0 Total)
MAKESEC:Security file installed as /HCL/HWA/WA/TWSDATA/Security
[wauser@svcas0246 TWS]$ ▓
```

Figure 16-10. *dumpsec Command*

- We have successfully given access to the users "HWA_Sched and HWA_operator" using traditional model.

Note Usually in production, LDAP or OpenID Connect is used for user authentication; in that case, there is no need to add the users to the xml files.

Role-Based Access Control

In HWA, we can give role-based security access as well. We have to define the role-based security model in the master domain manager database by using the Manage Workload Security interface from Dynamic Workload Console or the composer command-line utility. The default configuration to use role-based value "enRoleBasedSecurityFileCreation" is set to "yes," which means that the role-based security model is enabled after the master domain manager installation. In order to check if the role-based access control is enabled, run the command as shown in Figure 16-11.

```
[wauser@svcas0246 TWS]$ optman ls | grep -i role
enRoleBasedSecurityFileCreation / rs = YES
```

Figure 16-11. *enRoleBasedSecurityFileCreation Option*

Once role-based access is enabled, create the users using role-based access control. Please see the following requirements:

1. Create cust1,cust2 users and add them to group "customer." cust1 and cust2 users should have operator, analyst, and developer roles, and they should manage HWA objects created on "C1" folder only.

2. Create tenant1, tenant2 users and add them to "Tenant" group. tenant1 and tenan2 users should have operator, analyst, and developer roles, and they should manage HWA objects created on "C2" folder only.

A folder is used to organize the workload objects into different categories.

- We are using keycloak for access management, so we should add cust1&2 and tenant1&2 users on their respective groups. Log in to keycloak and add the users into respective groups. Refer to Figure 16-12 and 16-13.

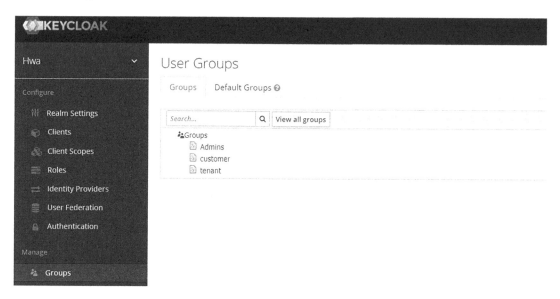

Figure 16-12.   Groups Added into Keycloak

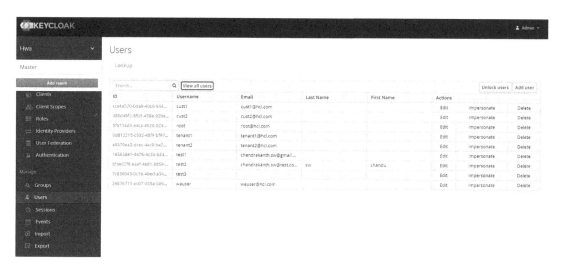

Figure 16-13.   Users Added into Groups

- Once the users are added into keycloak, add the groups into Dynamic Workload Console. To add the groups in DWC, navigate to Administration ➤ Manage role ➤ operator. Refer to Figure 16-14.

Figure 16-14. *Groups Added into Roles*

- Perform the same steps to add groups on analyst and developer roles, and create the security role, domain, and access control list using composer utility command as the following. Refer to Figures 16-15 to 16-20.

```
Folder creation:

FOLDER /C1/ END
FOLDER /C1/ END
FOLDER /C1/C1W/ END
FOLDER /SVC/ END
FOLDER /SVC/C1/ END
FOLDER /SVC/C1/ END

Dynamic Pool for the C1 Tenant:

CPUNAME C1_CLOUD
  DESCRIPTION "Cloud Pool Workstation"
  OS OTHER
  FOR MAESTRO HOST HWA_DWB
    TYPE POOL
  MEMBERS
    /C1/C1L_HWA_AGENT
END
```

Figure 16-15. *Folder Creation*

```
Role Based Security Domains:

SECURITYDOMAIN C1_POOL
  DESCRIPTION "Security Domain for C1 Pool"
  JOB CPU="C1_CLOUD"+FOLDER="/C1/"
  RESOURCE CPU="HWA_DWB"+FOLDER="/C1/"
  CPU CPU="C1_CLOUD"+TYPE="POOL"
END

SECURITYDOMAIN C1_SERVICE
  DESCRIPTION "Security Domain for C1 Service"
  JOB CPU="HWA_VMDM"+FOLDER="/SVC/C1/"
  SCHEDULE CPU="HWA_VMDM"+FOLDER="/SVC/C1/"
  EVENTRULE FOLDER="/SVC/C1/"
  VARTABLE FOLDER="/SVC/C1/"~NAME="$DEFAULT"
  FOLDER NAME="/SVC/C1/"
END
SECURITYDOMAIN C1_CPU
  DESCRIPTION "Security Domain for C1 Pool and Cpuclass CPUs"
  CPU FOLDER="/C1/"+TYPE="POOL","CPUCLASS"
END
SECURITYDOMAIN C1_SECURITY
  DESCRIPTION "Security Domain for objects not on folders"
  USEROBJ CPU="C1@"
  JOB CPU="C1@"+FOLDER="/C1/"~CPU="C1_POOL"
  SCHEDULE CPU="C1@"+FOLDER="/C1/"
  RESOURCE CPU="C1@"+FOLDER="/C1/"
  CPU CPU="C1@"~CPU="C1_POOL"
  PARAMETER CPU="C1@"
  ACTION PROVIDER="TWSACTION"
  EVENT PROVIDER="FILEMONITOR","TWSOBJECTSMONITOR"
  LOB NAME="C1"
END
SECURITYDOMAIN C1_JOBS_POOL
  DESCRIPTION "Security Domain for working on JOBS jobstream on /C1/C1L_HWA_AGENT"
  JOB CPU="/C1/C1L_HWA_AGENT"
END

SECURITYDOMAIN C1_JOBS
  DESCRIPTION "Security Domain for working on JOBS jobstream"
  JOB CPU="C1@"~CPU="C1_POOL"
  JOB CPUFOLDER="/C1/"
  SCHEDULE CPU="C1@"+NAME="JOBS"
  SCHEDULE NAME="JOBS"+CPUFOLDER="/C1/"
END
SECURITYDOMAIN C1_ADMIN
  DESCRIPTION "Admin Domain for C1"
  FILE NAME="SECURITY_DOM_C1@"
END
```

Figure 16-16. *Security Domain Creation*

```
Role Based Security role:

SECURITYROLE SECROLE_C1_ADMIN
  DESCRIPTION "Admin Security Role"
  FILE DISPLAY
  FOLDER ADD,DELETE,DISPLAY,MODIFY,USE,LIST,UNLOCK,ACL
END

SECURITYROLE SECROLE_C1_ANYUSER
  DESCRIPTION "AnyUser Security Role"
  CPU DISPLAY,LIST
  CALENDAR DISPLAY,USE,LIST
  FILE DISPLAY
END

SECURITYROLE SECROLE_C1_CPU
  DESCRIPTION "Pool CPU Security Role"
  CPU ADD,CONSOLE,DELETE,DISPLAY,FENCE,LIMIT,LINK,MODIFY,SHUTDOWN,START,STOP,UNLINK,USE,LIST,UNLOCK,RUN,RESETFTA,MANAGE
END

SECURITYROLE SECROLE_C1_FULLACCESS
  DESCRIPTION "FullAccess Security Role"
  JOB ADD,ADDDEP,ALTPRI,CANCEL,CONFIRM,DELDEP,DELETE,DISPLAY,KILL,MODIFY,RELEASE,REPLY,RERUN,SUBMIT,USE,LIST,UNLOCK,SUBMITDB,RUN
  SCHEDULE ADD,ADDDEP,ALTPRI,CANCEL,DELDEP,DELETE,DISPLAY,LIMIT,MODIFY,RELEASE,REPLY,SUBMIT,LIST,UNLOCK
  CPU CONSOLE,DELETE,DISPLAY,FENCE,LIMIT,LINK,MODIFY,SHUTDOWN,START,STOP,UNLINK,USE,LIST,UNLOCK,RUN,RESETFTA,MANAGE
  RESOURCE ADD,DELETE,DISPLAY,MODIFY,RESOURCE,USE,LIST,UNLOCK
  PROMPT ADD,DELETE,DISPLAY,MODIFY,REPLY,USE,LIST,UNLOCK
  CALENDAR ADD,DELETE,DISPLAY,MODIFY,USE,LIST,UNLOCK
  PARAMETER ADD,DELETE,DISPLAY,MODIFY,LIST,UNLOCK
  REPORT DISPLAY
  RUNCYGRP ADD,DELETE,DISPLAY,MODIFY,USE,LIST,UNLOCK
  USEROBJ ADD,DELETE,DISPLAY,MODIFY,USE,ALTPASS,LIST,UNLOCK
  VARTABLE ADD,DELETE,DISPLAY,MODIFY,USE,LIST,UNLOCK
  ACTION DISPLAY,SUBMIT,USE,LIST
  EVENT USE
  EVENTRULE ADD,DELETE,DISPLAY,MODIFY,LIST,UNLOCK
  WKLDAPPL ADD,DELETE,DISPLAY,MODIFY,LIST,UNLOCK
  LOB USE
  FOLDER ADD,DELETE,DISPLAY,MODIFY,USE,LIST,UNLOCK
END
```

Figure 16-17. *Security Role Creation*

```
SECURITYROLE SECROLE_C1_JOBS
  DESCRIPTION "JOBS Security Role"
  JOB ADDDEP,ALTPRI,CANCEL,CONFIRM,DELDEP,DISPLAY,KILL,RELEASE,REPLY,RERUN,LIST,SUBMITDB,SUBMIT
  SCHEDULE ADDDEP,ALTPRI,CANCEL,DELDEP,DISPLAY,LIMIT,RELEASE,REPLY,LIST
END

SECURITYROLE SECROLE_C1_JOBS_POOL
  DESCRIPTION "JOBS on Pool Security Role"
  JOB ADDDEP,ALTPRI,CANCEL,CONFIRM,DELDEP,DISPLAY,KILL,RELEASE,REPLY,RERUN,LIST,SUBMITDB
END

SECURITYROLE SECROLE_C1_OPERATOR
  DESCRIPTION "Operator Security Role"
  JOB ADDDEP,ALTPRI,CANCEL,CONFIRM,DELDEP,DISPLAY,KILL,RELEASE,REPLY,RERUN,LIST,SUBMITDB,RUN
  SCHEDULE ADDDEP,ALTPRI,CANCEL,DELDEP,DISPLAY,LIMIT,RELEASE,REPLY,SUBMIT,LIST
  CPU CONSOLE,DISPLAY,FENCE,LIMIT,LINK,SHUTDOWN,START,STOP,UNLINK,LIST,RUN,RESETFTA,MANAGE,USE
  RESOURCE DISPLAY,RESOURCE,USE,LIST
  PROMPT DISPLAY,LIST
  CALENDAR DISPLAY,LIST
  PARAMETER DISPLAY,LIST
  REPORT DISPLAY
  RUNCYGRP DISPLAY,LIST
  VARTABLE DISPLAY,LIST
  ACTION DISPLAY,LIST
  EVENT USE
  EVENTRULE DISPLAY,LIST
  WKLDAPPL DISPLAY,LIST
  LOB USE
  FOLDER DISPLAY,LIST
END

SECURITYROLE SECROLE_C1_POOL
  DESCRIPTION "Pool Security Role"
  JOB ADD,ADDDEP,ALTPRI,CANCEL,CONFIRM,DELDEP,DELETE,DISPLAY,KILL,MODIFY,RELEASE,REPLY,RERUN,USE,LIST,UNLOCK,SUBMITDB,RUN
  CPU CONSOLE,DISPLAY,FENCE,LIMIT,LINK,SHUTDOWN,START,STOP,UNLINK,LIST,UNLOCK,RUN,RESETFTA,MANAGE,USE
  RESOURCE ADD,DELETE,DISPLAY,MODIFY,RESOURCE,USE,LIST,UNLOCK
END
```

Figure 16-18. *Security Role Creation*

```
SECURITYROLE SECROLE_C1_SCHEDULER
  DESCRIPTION "Scheduler Security Role"
  JOB ADD,DELETE,DISPLAY,MODIFY,USE,LIST,UNLOCK,RUN
  SCHEDULE ADD,DELETE,DISPLAY,LIMIT,MODIFY,LIST,UNLOCK
  CPU CONSOLE,DELETE,DISPLAY,FENCE,LIMIT,LINK,MANAGE,MODIFY,SHUTDOWN,START,STOP,UNLINK,LIST,UNLOCK,RUN,USE
  RESOURCE ADD,DELETE,DISPLAY,MODIFY,USE,LIST,UNLOCK
  PROMPT ADD,DELETE,DISPLAY,MODIFY,USE,LIST,UNLOCK
  CALENDAR ADD,DELETE,DISPLAY,MODIFY,USE,LIST,UNLOCK
  PARAMETER ADD,DELETE,DISPLAY,MODIFY,LIST,UNLOCK
  REPORT DISPLAY
  RUNCYGRP ADD,DELETE,DISPLAY,MODIFY,USE,LIST,UNLOCK
  VARTABLE ADD,DELETE,DISPLAY,MODIFY,USE,LIST,UNLOCK
  ACTION DISPLAY,SUBMIT,USE,LIST
  EVENT USE
  EVENTRULE ADD,DELETE,DISPLAY,MODIFY,LIST,UNLOCK
  WKLDAPPL ADD,DELETE,DISPLAY,MODIFY,LIST,UNLOCK
  LOB USE
  FOLDER ADD,DELETE,DISPLAY,MODIFY,USE,LIST,UNLOCK
END
SECURITYROLE SECROLE_C1_SERVICE
  DESCRIPTION "Service Security Role"
  JOB ADDDEP,ALTPRI,CANCEL,CONFIRM,DELDEP,DISPLAY,KILL,RELEASE,REPLY,RERUN,LIST,SUBMITDB,RUN
  SCHEDULE ADDDEP,ALTPRI,CANCEL,DELDEP,DISPLAY,LIMIT,RELEASE,REPLY,SUBMIT,LIST
  VARTABLE DISPLAY,MODIFY,LIST,UNLOCK
  FOLDER ADD,DISPLAY,LIST,USE
END
SECURITYROLE SECROLE_C1_VIEWER
  DESCRIPTION "Viewer Security Role"
  JOB DISPLAY,LIST
  SCHEDULE DISPLAY,LIST
  CPU DISPLAY,LIST,USE
  RESOURCE DISPLAY,LIST,USE
  PROMPT DISPLAY,LIST,USE
  CALENDAR DISPLAY,LIST,USE
  PARAMETER DISPLAY,LIST
  REPORT DISPLAY
  RUNCYGRP DISPLAY,LIST,USE
  VARTABLE DISPLAY,LIST,USE
  ACTION DISPLAY,LIST
  EVENT USE
  EVENTRULE DISPLAY,LIST
  WKLDAPPL DISPLAY,LIST
  LOB USE
  FOLDER DISPLAY,LIST
END
```

Figure 16-19. *Security Role Creation*

- Perform the preceding same steps for tenant group. Once the RBAC is created on master domain manager for both the groups, we can log in to cust 1 and tenant 1 users to check the access. Log in to Dynamic Workload Console using cust1 and validate if cust1 user is able to see the objects under "C1" folder. Refer to Figure 16-21.

```
Role Based ACL:

ACCESSCONTROLLIST FOLDER /C1
  group: customer SECROLE_C1_FULLACCESS ,SECROLE_C1_ADMIN
END

ACCESSCONTROLLIST FOR C1_ADMIN
  group: customer SECROLE_C1_ADMIN
END

ACCESSCONTROLLIST FOR C1_JOBS
  group: customer SECROLE_C1_JOBS
END

ACCESSCONTROLLIST FOR C1_JOBS_POOL
  group: customer SECROLE_C1_JOBS_POOL
END

ACCESSCONTROLLIST FOR C1_POOL
  group: customer SECROLE_C1_POOL
END

ACCESSCONTROLLIST FOR C1_SECURITY
  group: customer SECROLE_C1_FULLACCESS
END

ACCESSCONTROLLIST FOR C1_SERVICE
  group: customer SECROLE_C1_SERVICE
END
```

Figure 16-20. *Access Control Creation*

Figure 16-21. *cust1 User Console View*

- We can see that cust1 user is able to see only the objects under "C1" folder. Log in to tenant 1 and validate if the user can see the objects under "C2" folder only. Refer to Figure 16-22.

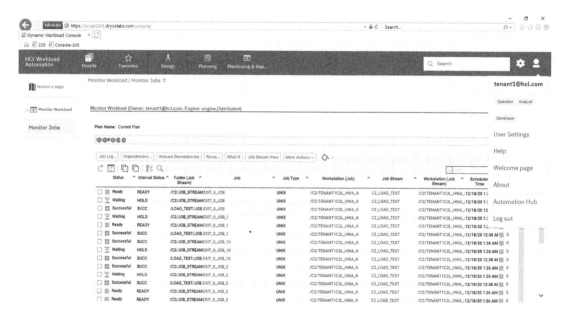

Figure 16-22. *tenant1 User Console View*

- We can see that tenant1 user is able to see only the objects under "C2" folder.

- We have successfully given access to cust1&2 and tenant1&2 users using role-based access control.

Summary

In this chapter, we covered the following capabilities of HWA:

- Access creation for users in traditional security model
- Access creation for users/groups using role-based access control

Index

A

Alerting mechanism
>email notifications, 191
>end users, 185
>event rule, 186–190
>job failure email notification, 185
>job status, 188
>restore outages, 191

Ansible
>definition, 153
>inventory file, 153
>requirements
>>action tab, 157
>>configure job, 153
>>DWC home page, 154
>>job properties, 156
>>plug-in view, 155
>>schedule job, 159
>>submit job, 158
>yml file, 153

Application monitoring event, 188

B

Business Intelligence and Reporting
>Tool (BIRT), 55, 208

C

Command-line interface (CLI), 12, 49, 211

D

Daily health check, 173
Disaster Recovery (DR), 5, 6, 23–24, 35–39
dumpsec command, 211, 215, 216
Dynamic agent (DA), 9, 15–16, 19, 97, 111,
>126, 133, 141, 147, 156
Dynamic pool (DP), 9–10, 15, 16
Dynamic Workload Broker (DWB), 16
Dynamic Workload Console (DWC),
>14, 27
>ad hoc jobs, 53
>business requirements, 55
>>actions view, 95
>>Ad Hoc job, 89–91
>>dependency options, 71, 72
>>dependency selection, 78–81
>>design job, 56
>>HWA agent, 89
>>job log, 87
>>job properties, 58–60
>>jobs, 56
>>job selection, 67, 68
>>job status, 92, 94, 95
>>job stream, 61–63, 69, 70, 73, 74, 77,
>>78, 81, 87, 88
>>monitoring job, 92
>>object-type selection, 86
>>rerun options, 93, 94
>>run cycle, 64–66, 75, 76
>>submit job, 82–85

© Navin Sabharwal and Subramani Kasiviswanathan 2023
N. Sabharwal and S. Kasiviswanathan, *Workload Automation Using HWA*,
https://doi.org/10.1007/978-1-4842-8885-6

Dynamic Workload
    Console (DWC) (*cont.*)
  CLI, 49
  HWA login page, 50
  landing page, 52
  predefined job, 53
  tabs
    administration, 51
    design, 51, 52
    forecast plan, 52
    monitoring/reporting, 53–55
    planning, 52
    preproduction plan, 53
    trial plan, 52
    workload submission, 53
  web browser, 49

E

"exit 1" command, 166
Explicit File Transfer Protocol (FTPES), 97
Extended agent (XA), 10, 19

F

Fault tolerant agent (FTA), 8, 9, 14, 15, 17,
    19, 37, 38, 97, 111, 133, 156
File monitoring event, 188
File Transfer Protocol (FTP), 97, 98,
    100, 107
File Transfer Protocol Secure (FTPS), 97
File transfers
  business processes, 98
    connection properties, job, 102
    destination file information,
      105, 106
    destination server files, 110
    FTP options, 107

    FTP plug-in, 100
    job created, 108
    job definition, 100, 101
    job submission, 108
    log information, 109
    source file information, 104
    SSH, 98
    test connection results, 103
    workflow, 97, 98
  destination server, 97
  DWC home page, 99
  security protocols, 97

G

GitHub, 48

H, I

HCL workload automation (HWA),
    1, 49
  advantages, 11, 12
  application HA capability, 14
  architecture types, 21–24
  audit, 14
  communication ports, 20
  components, 6, 7, 9, 10, 13–16
  console, 7
  DA, 15
  database, 16
  definition, 6
  DWB, 16
  FTA, 15
  HWA master, 13
  interprocess communication, 17
  JnextPlan, 11
  KPIs, 6
  liberty, 16

messaging communication, 17–19

multiple deployment options (*see* Planning deployments)

network, 8

Process Status Command, 193

reporting

    personalized, 205

    predefined, 205

scheduler, 13

software, 8

terminologies, 10

types, 205

HWA event rule management

auto-remediate job failures

    agent running, 171

    batchman process, 167–169, 171, 172

    event rule properties, 169

    event status, 170

    HWA process, 167

    server log file, 171

ServiceNow, 172

    actions, 165, 166

    add event options, 163

    create event rule, 162

    definition, 161

    event parameters, 164

    job failure ticket, 162

    job status, 164, 166

xml format, 161

Hype cycle for IT operations management, 1

J

JNextPlan, 10, 11, 173, 174

Job scheduling software, 2

K

Kubernetes cluster

HWA capabilities, 144

    batch jobs, 141

    cluster config file information, 143

    connection parameters, 142

    job definition properies, 142

    job plug-in, 141

    mysql-deployment, 139

    process information, 143

    submit job, 144

HWA dynamic agent, 139

L

Liberty, 10, 16, 21, 27, 32, 33, 35, 36, 191, 192, 196, 197, 203

Liberty Process Stop command, 192

M, N

makesec command, 211, 216

Master domain manager (MDM), 8–10, 19–22, 27, 38, 56, 167, 173, 217, 224

MDM_DA agent server, 98

Microsoft Azure

definition, 145

VM, 145

    client/tenant, 148

    connection parameters, 149

    create new VM, 151

    DWC home page, 145

    exiting VM options, 150

    job definition, 147

    plug-in, 147

    submit job, 151, 152

Microsoft SQL server
 definition, 123
 HWA integration, 129
 use cases, 123
  database selection, 127
  DWC home page, 124
  HWA job properties, 126
  job selection, 128
  KPI, 123
  plug-in, 125
  submit job options, 129

O

"optman ls" command, 195
Outages
 db connectivity check, 195
 DWC liberty status, 197, 198
 DWC process start commands, 197
 HWA process status command, 193
 HWA process stop commands, 193
 HWA start commands, 196
 liberty process, 192
 planned, 191
 startup command, 194
 types, 191
 unplanned/emergency, 194

P, Q

Personalized reports
 BIRT, 208
 HWA Master, 208
 output, 209
 report parameters, 209
Planning deployments
 containerized deployment, 39–47
 disaster recovery, 35–39

 high availability architecture, 32–35
 HWA deployment, 47, 48
 memory requirements, 26
 prerequisites, 25
 stand-alone architecture, 27–32
Pool, 9, 15, 16
Predefined reports
 definition, 205
 filter, 207
 header, 206
 name page, 208
 output format, 207, 208
 reporting page, 206
 types, 205

R

Relational database management
  system (RDBMS), 19, 123
REST API
 definition, 131
 HTTP methods, 131
 HWA, 131
 use case
  authentication, 135
  DWC home page, 132
  GET, 132
  HWA rest api job, 136, 137
  job properties
   selection, 134
  methods, 135
  plug-in page, 133
  service URL, 138
  submit job, 137
  web URL, 132
"rmstdlist" script, 176, 178
Role-based access control (RBAC), 14
 access role creation, 225

add groups, 219, 220

enRoleBasedSecurityFileCreation, 217

keycloak, 218, 219

requirements, 217

security domain creation, 221

security role creation, 222–224

tenenat1 user control view, 226

user control view, 226

S

SAP R3batch access method, 111

SAP-related event, 188

Scheduler, 2, 3, 5, 6, 13, 51, 111, 191

script/command, 53, 153

Secure Shell (SSH), 15, 97, 98, 101, 156

Security file

role-based access control, 211

traditional model (file based), 211

ServiceNow, 161–167

Sinfonia, 11

Stageman, 11

Stand-alone architecture, 21, 27–33, 35, 36, 39, 47

Standard agent (SA), 10, 19

StartAppServer, 10, 196, 203

StartUp Command, 194, 195

SymNew, 10, 11

Symphony, 10, 11, 37

Systems Applications and Product (SAP)

ABAP jobs, 111

business processes, 111

design page, 116

job status, 122

job submission, 120

library files, 113, 114

monitoring workload, 121

plug-in, 117

properties, 118, 119

R3batch options, 114, 115

workflow, 112

enterprise scheduler, 111

plug-in, 122

products, 111

T

Time-based and event-driven scheduling, 3

Tool administration

daily health check, 173–175

database backups/restore, 183

database maintenance, 181–183

housekeeping maintains application, 176, 181

DWC home page, 177

HWA job properties, 178, 179

HWA Unix job, 180, 181

log files, 176

Unix job, 177

Traditional model (file based)

authentication file, 215

developer roles, 214, 215

DWC, 211, 212

enRoleBasedSecurityFileCreation, 211

managed roles, 212, 213

modifying file, 216

Troubleshooting techniques

agent status, 199–201

DWC, 203

linked agent, 202

monitor workload page, 199

telnet connection, 202

use cases, 198

U, V

Unplanned/emergency outages, 194

W, X, Y, Z

Workload automation
    application-level high availability, 5
    application platform/integration, 3
    architecture, 5
    batch processing, 4
    batch/workflow, 5
    disaster recovery, 6

    events/errors, 5
    features, 1
    handle/store outputs, 4
    job stream, 1–3
    requirements, 1, 2
    security, 6
    self-service capabilities, 5
    virtual/physical resources, 4

add groups, 219, 220

enRoleBasedSecurityFileCreation, 217

keycloak, 218, 219

requirements, 217

security domain creation, 221

security role creation, 222–224

tenenat1 user control view, 226

user control view, 226

S

SAP R3batch access method, 111

SAP-related event, 188

Scheduler, 2, 3, 5, 6, 13, 51, 111, 191

script/command, 53, 153

Secure Shell (SSH), 15, 97, 98, 101, 156

Security file

    role-based access control, 211

    traditional model (file based), 211

ServiceNow, 161–167

Sinfonia, 11

Stageman, 11

Stand-alone architecture, 21, 27–33, 35, 36, 39, 47

Standard agent (SA), 10, 19

StartAppServer, 10, 196, 203

StartUp Command, 194, 195

SymNew, 10, 11

Symphony, 10, 11, 37

Systems Applications and Product (SAP)

    ABAP jobs, 111

    business processes, 111

        design page, 116

        job status, 122

        job submission, 120

        library files, 113, 114

        monitoring workload, 121

        plug-in, 117

        properties, 118, 119

        R3batch options, 114, 115

        workflow, 112

    enterprise scheduler, 111

    plug-in, 122

    products, 111

T

Time-based and event-driven scheduling, 3

Tool administration

    daily health check, 173–175

    database backups/restore, 183

    database maintenance, 181–183

    housekeeping maintains application, 176, 181

        DWC home page, 177

        HWA job properties, 178, 179

        HWA Unix job, 180, 181

        log files, 176

        Unix job, 177

Traditional model (file based)

    authentication file, 215

    developer roles, 214, 215

    DWC, 211, 212

    enRoleBasedSecurityFileCreation, 211

    managed roles, 212, 213

    modifying file, 216

Troubleshooting techniques

    agent status, 199–201

    DWC, 203

    linked agent, 202

    monitor workload page, 199

    telnet connection, 202

    use cases, 198

U, V

Unplanned/emergency outages, 194

W, X, Y, Z

Workload automation
    application-level high availability, 5
    application platform/integration, 3
    architecture, 5
    batch processing, 4
    batch/workflow, 5
    disaster recovery, 6
    events/errors, 5
    features, 1
    handle/store outputs, 4
    job stream, 1–3
    requirements, 1, 2
    security, 6
    self-service capabilities, 5
    virtual/physical resources, 4

add groups, 219, 220

enRoleBasedSecurityFileCreation, 217

keycloak, 218, 219

requirements, 217

security domain creation, 221

security role creation, 222–224

tenenat1 user control view, 226

user control view, 226

S

SAP R3batch access method, 111

SAP-related event, 188

Scheduler, 2, 3, 5, 6, 13, 51, 111, 191

script/command, 53, 153

Secure Shell (SSH), 15, 97, 98, 101, 156

Security file

    role-based access control, 211

    traditional model (file based), 211

ServiceNow, 161–167

Sinfonia, 11

Stageman, 11

Stand-alone architecture, 21, 27–33, 35,
      36, 39, 47

Standard agent (SA), 10, 19

StartAppServer, 10, 196, 203

StartUp Command, 194, 195

SymNew, 10, 11

Symphony, 10, 11, 37

Systems Applications and Product (SAP)

    ABAP jobs, 111

    business processes, 111

      design page, 116

      job status, 122

      job submission, 120

      library files, 113, 114

      monitoring workload, 121

      plug-in, 117

      properties, 118, 119

    R3batch options, 114, 115

    workflow, 112

enterprise scheduler, 111

plug-in, 122

products, 111

T

Time-based and event-driven
      scheduling, 3

Tool administration

    daily health check, 173–175

    database backups/restore, 183

    database maintenance, 181–183

    housekeeping maintains application,
      176, 181

      DWC home page, 177

      HWA job properties, 178, 179

      HWA Unix job, 180, 181

      log files, 176

      Unix job, 177

Traditional model (file based)

    authentication file, 215

    developer roles, 214, 215

    DWC, 211, 212

    enRoleBasedSecurityFileCreation, 211

    managed roles, 212, 213

    modifying file, 216

Troubleshooting techniques

    agent status, 199–201

    DWC, 203

    linked agent, 202

    monitor workload page, 199

    telnet connection, 202

    use cases, 198

U, V

Unplanned/emergency outages, 194

W, X, Y, Z

Workload automation
  application-level high availability, 5
  application platform/integration, 3
  architecture, 5
  batch processing, 4
  batch/workflow, 5
  disaster recovery, 6
  events/errors, 5
  features, 1
  handle/store outputs, 4
  job stream, 1–3
  requirements, 1, 2
  security, 6
  self-service capabilities, 5
  virtual/physical resources, 4

Printed in the United States
by Baker & Taylor Publisher Services